T0290952

Guide to Electrical Power Distribution Systems

Sixth Edition

Guide to Electrical Power Distribution Systems

Sixth Edition

Anthony J. Pansini, EE, PE
Life Fellow IEEE, Sr. Member ASTM

Library of Congress Cataloging-in-Publication Data

Pansini, Anthony J.
Guide to electrical power distribution systems/Anthony J. Pansini.--6th ed.
p. cm.
Includes index.
ISBN: 0-88173-505-1 (print) — 0-88173-506-X (electronic)
1. Electric power distribution. 2. Electric power transmission. I. Title.

[TK3001.P284 2005]
621.319--dc22

2004056257

Guide to electrical power distribution systems, sixth edition/Anthony J. Pansini

Published by The Fairmont Press, Inc.
700 Indian Trail
Lilburn, GA 30047
tel: 770-925-9388; fax: 770-381-9865
http://www.fairmontpress.com

Distributed by Marcel Dekker/CRC Press
2000 N.W. Corporate Blvd.
Boca Raton, FL 33431
tel: 800-272-7737
http://www.crcpress.com

Printed in the United States of America
10 9 8 7 6 5 4 3 2 1

0-88173-505-1 (The Fairmont Press, Inc.)
0-8493-3666-X (Dekker/CRC Press)

To my superiors, associates, colleagues, and subordinates, from whom–along the way–I learned much

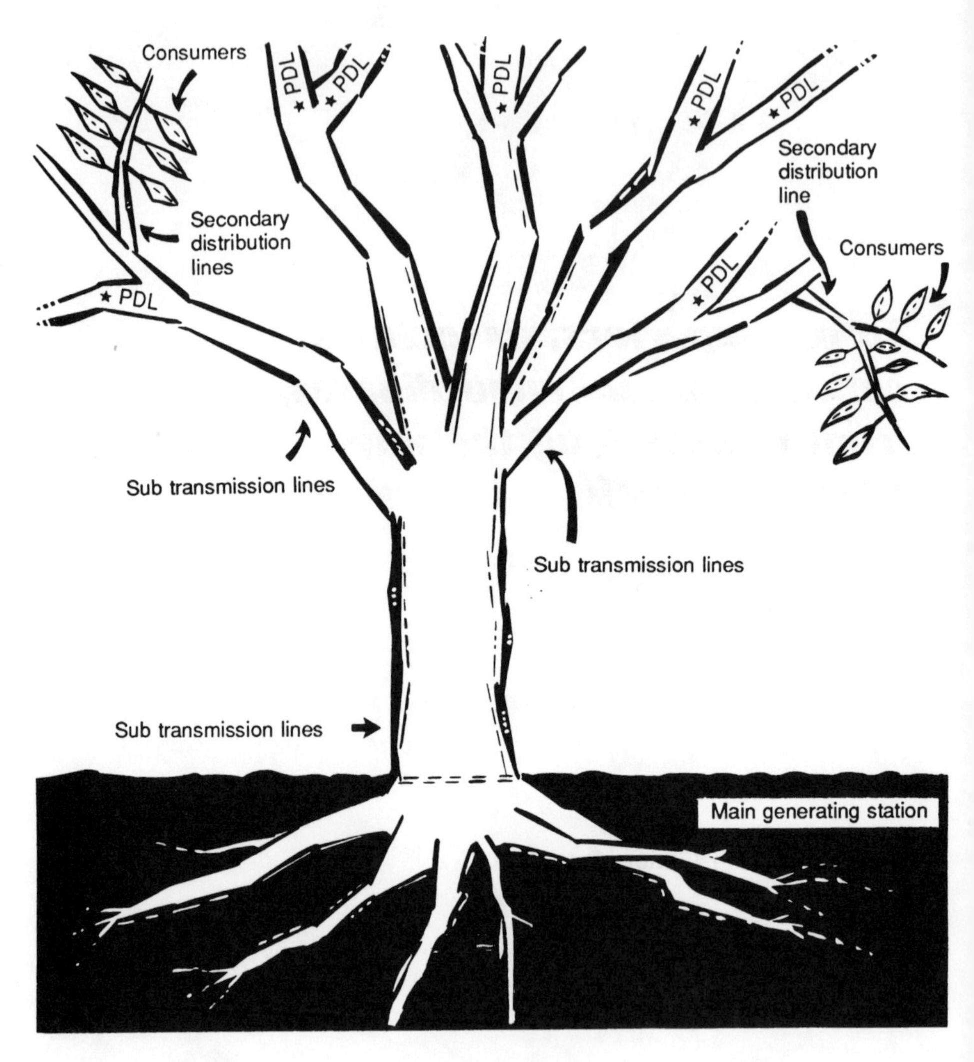

Tree of Electricity

Transmission and Distribution
PDL: Primary distribution lines, ★ indicates location of transformer stations.

Contents

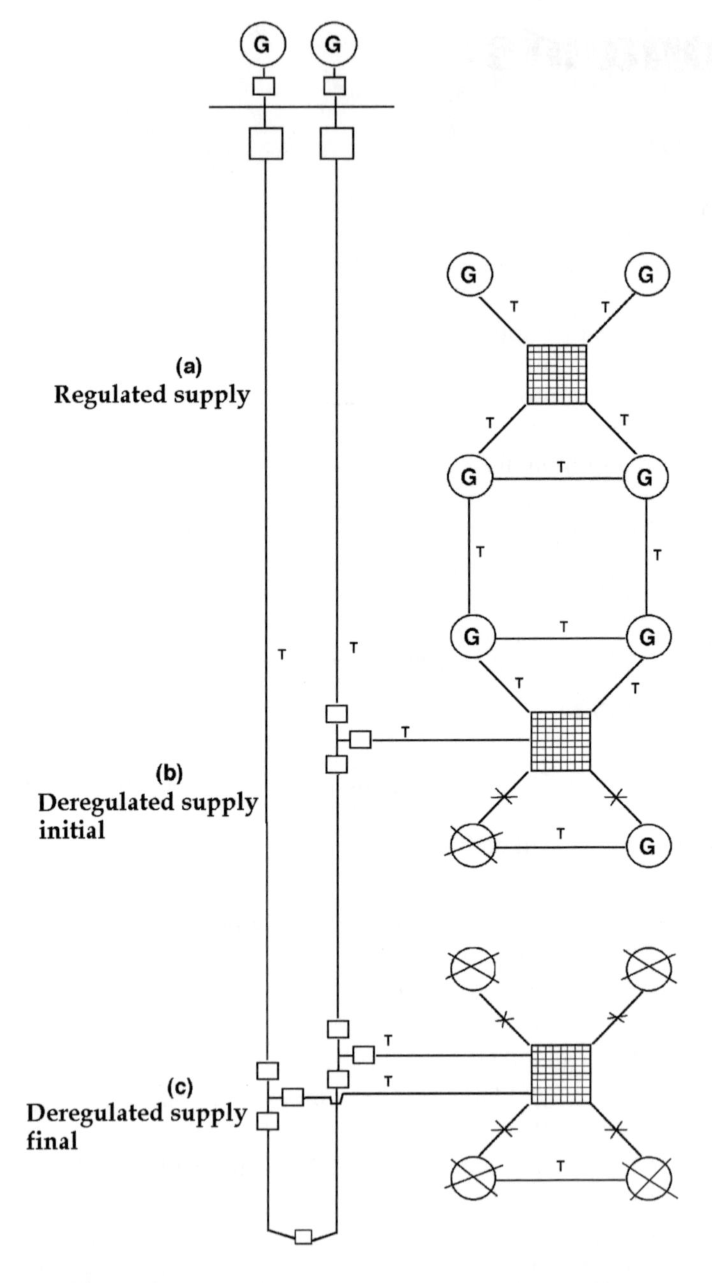

Figure P-1. Simplified schematic diagram of transition from regulated to deregulated supply systems.

Preface to the Sixth Edition

This edition continues the practice of updating its contents to reflect changes affecting electric distribution systems. It continues its original role of providing information in a non or semi technical manner to persons working on such systems enabling them to perform their duties in an enlightened way. This contributes not only to a greater quality of production, but more importantly, with greater safety to themselves and the general public, the consumers they serve.

While the effects of deregulation have been more pronounced on the generation and transmission components of the supply chain, it has also had an impact on the distribution system with its cogeneration and distributed generation features.

Improvement in materials and methods continue to contribute to the economic and environmental betterment enjoyed by consumers. Predominant among these include the gradual replacement of heavier porcelain insulators with polymer (plastic) ones, the employment of insulated bucket vehicles making climbing with the use of spikes a lost art. Improvements in "solid" type insulation in cables, and more efficient machinery for placing them underground. Thus narrowing the economic differences between overhead and underground installations.

The distribution system is the most visible part of the supply chain, and as such the most exposed to the critical observation of its users. It is, in many cases, the largest investment, maintenance and operation expense, and the object of interest to government, financial agencies, and "watch dog" associations of concerned citizens. As such, the desirability of knowing how and why it is so constituted becomes obvious.

Distribution systems have also been affected by deregulation, although not in the same manner as transmission systems, Figure P-1 (opposite). Where additional transmission or generation was not available or too great an expense to supply some additional loads, Distributed Generation made its entry on Distribution Systems. Here, small generating units usually powered by small gas turbines (although other units such as wind powered, solar, fuel cells, etc. may be involved) are connected to the system as are cogeneration units, both with possible hazard to safety.

These notes were begun in the early 1940's as classroom material, part of a rapid training program for line personnel. The program was highly successful, reflected in greater safety and production among other benefits, and achieved national attention.

Once again, our thanks to our old friends, Ken Smalling and The Fairmont Press for their help and support.

Anthony J. Pansini
Waco, Texas, 2004

Chapter 1

The Transmission and Distribution System

INTRODUCTION

Like any other industry, the electric power system may be thought of as consisting of three main divisions:

1. manufacture, production or generation, cogeneration,
2. delivery or transmission and distribution,
3. consumption.

The discussions in this book will be limited to the subject of electric distribution.

POWER TRANSMISSION

Figures 1-1 and 1-2 show a typical transmission and distribution system in both pictorial and block diagram forms. Although geographical difficulties, demand variances, and other reasons may make for minor differences in some transmission and distribution systems, the voltages chosen here are pretty typical. This is what happens to electricity between the generator and a home, office, store, or factory.

There are many definitions of transmission lines, distribution circuits, and substations specifying distinctions between them. However, none of these definitions is universally applicable. To give some idea of where one ends and the other begins: Transmission may be compared to bulk delivery of a commodity from factory to regional depots; subtransmission from the depot to central area warehouses; primary distribution from area warehouse to local wholesale vendors; secondary distribution from the vendors to local stores; services from store to consumer.

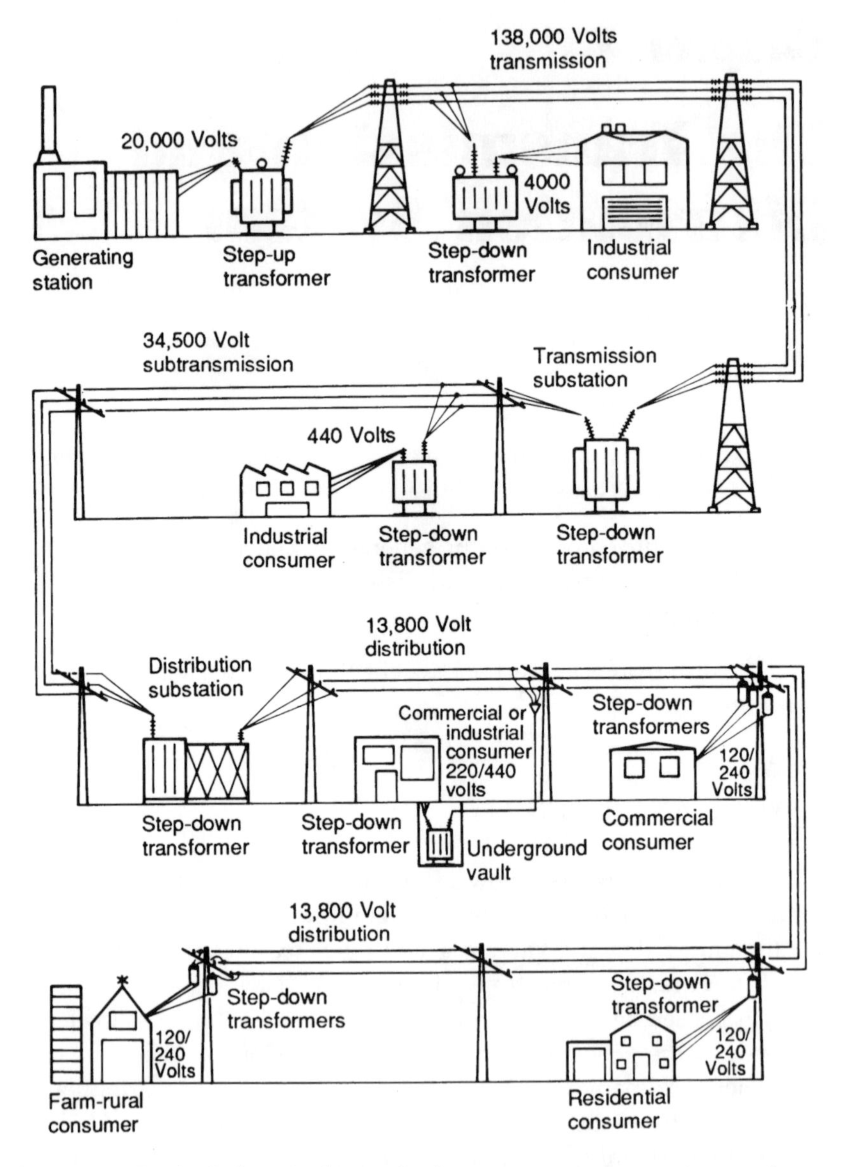

Figure 1-1. Typical electrical supply from generator to customer showing transformer applications and typical operating voltages.

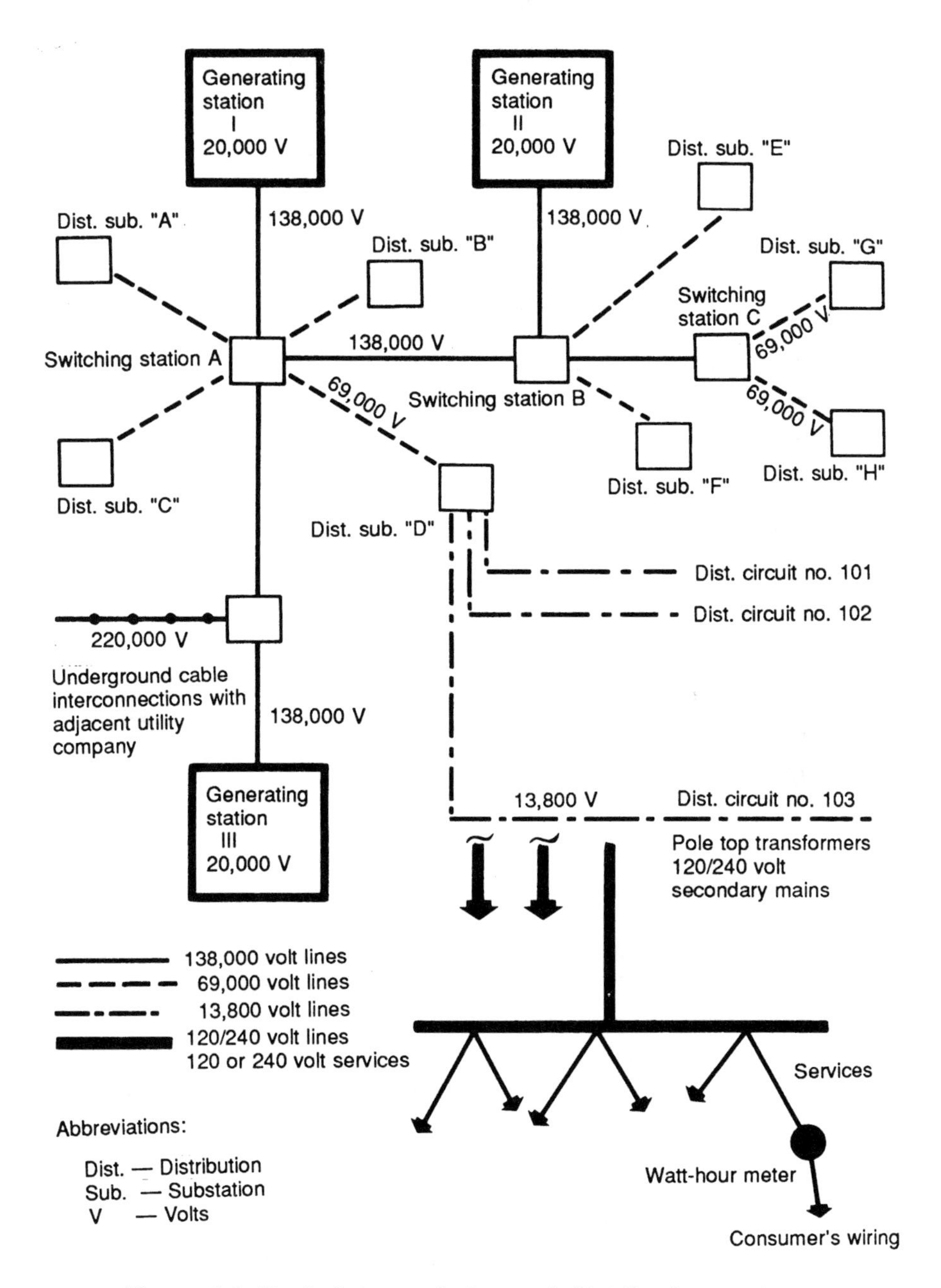

Figure 1-2. Typical transmission and distribution system.

In the pictorial rendition, note that the generator produces 20,000 volts. This, however, is raised to 138,000 volts for the long transmission journey. This power is conducted over 138,000-volt (138 kV) transmission lines to switching stations located in the important load area served. These steel tower or wood frame lines, which constitute the backbone of the transmission system, span fields and rivers in direct cross country routes. When the power reaches the switching stations, it is stepped down to 69,000 volts (69 kV) for transmission in smaller quantities to the substations in the local load areas. (In some cases it might be stepped down to 13,800 volts [13.8 kV] for direct distribution to local areas.) Transmission circuits of such voltages usually consist of open wires on wood or steel poles and structures in outlying zones (along highways, for example) where this type of construction is practicable.

Other transmission-line installations can provide an interchange of power between two or more utility companies to their mutual advantage. Sometimes, in more densely populated areas, portions of these transmission lines may consist of high-voltage underground systems operating at 69,000, 138,000, 220,000, 345,000, 500,000, and 750,000 volts.

WATER-CURRENT ANALOGY

The flow of electric current may be visualized by comparing it with the flow of water. Where water is made to flow in pipes, electric current is conducted along wires.

To move a definite amount of water from one point to another in a given amount of time, either a large-diameter pipe may be used and a low pressure applied on the water to force it through, or a small-diameter pipe may be used and a high pressure applied to the water to force it through. While doing this it must be borne in mind that when higher pressures are used, the pipes must have thicker walls to withstand that pressure (see Figure 1-3).

The same rule applies to the transmission of electric current. In this case, the diameter of the pipe corresponds to the diameter of the wire and the thickness of the pipe walls corresponds to the thickness of the insulation around the wire, as shown in Figure 1-4.

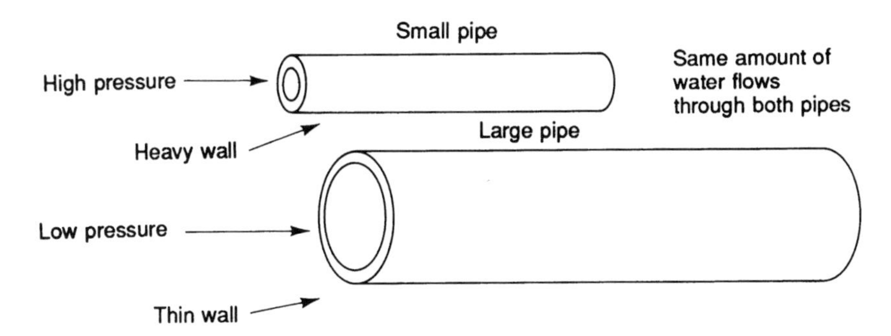

Figure 1-3. Comparison of water flow through different size pipes.

THE DISTRIBUTION SYSTEM

At the substations, the incoming power is lowered in voltage for distribution over the local area. Each substation feeds its local load area by means of *primary distribution feeders,* some operating at 2400 volts and others at 4160 volts and 13,800 volts or higher.

Ordinarily, primary feeders are one to five miles in length; in rural sections where demands for electricity are relatively light and scattered, they are sometimes as long as 10 or 12 miles. These circuits are usually

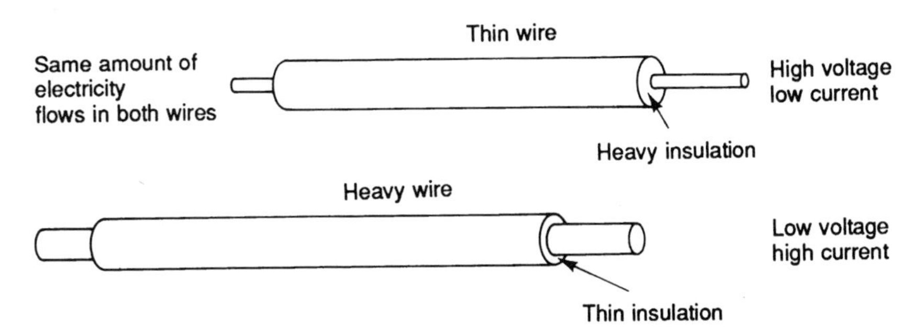

Figure 1-4. Comparison of current flow in different size wires.

carried on poles; but in the more densely built-up sections, underground conduits convey the cables, or the cable may be buried directly in the ground.

Distribution transformers connect to the primary distribution lines. These transformers step down the primary voltage from 2400 volts, 4160 volts, or 13,800 volts, as the case may be, to approximately 120 volts or 240 volts for distribution over secondary mains to the consumer's service (see Figure 1-5).

The lines which carry the energy at utilization voltage from the transformer to consumer's services are called secondary distribution mains and may be found overhead or underground. In the case of transformers supplying large amounts of electrical energy to individual consumers, no secondary mains are required. Such consumers are railroads, large stores, and factories. The service wires or cables are connected directly to these transformers. Transformers may also serve a number of

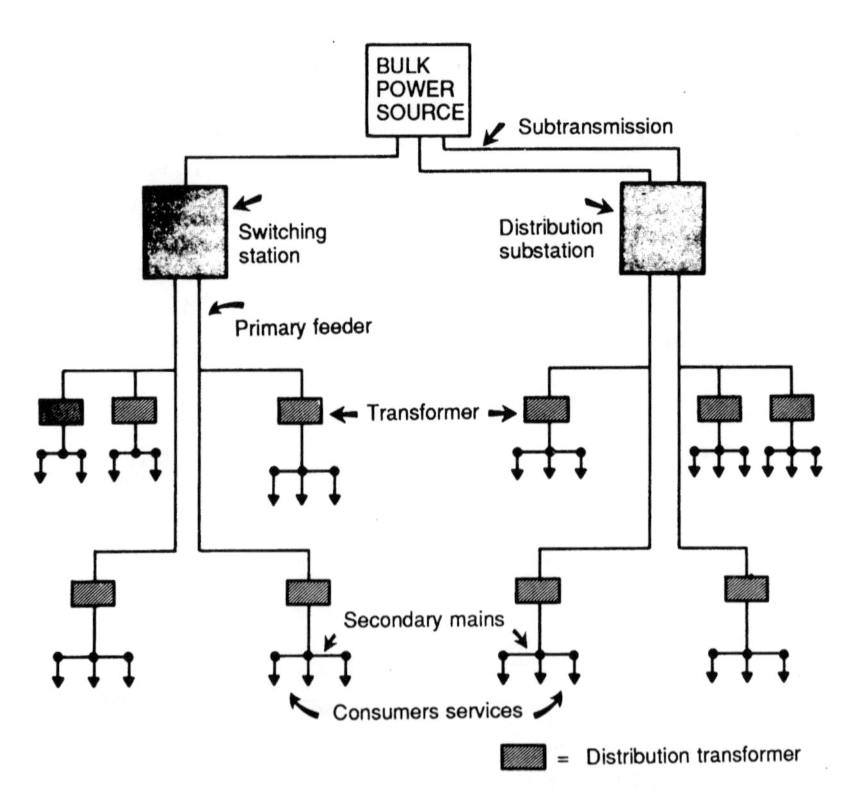

Figure 1-5. Typical distribution system showing component parts.

consumers and secondary mains; they are located in practically every street in the area served by utility companies.

Services and meters link the distribution system and the consumer's wiring. Energy is tapped from the secondary mains at the nearest location and carried by the service wires to the consumer's building. As it passes on to operate the lights, motors, and various appliances supplied by the house wiring, it is measured by a highly accurate device known as the watt-hour meter. The watt-hour meter represents the cash register of the utility company (see Figure 1-6).

DETERMINING DISTRIBUTION VOLTAGES

It was pointed out earlier that low voltages require large conductors, and high voltages require smaller conductors. This was illustrated with a water analogy. A small amount of pressure may be applied and the water will flow through a large pipe, or more pressure may be applied and the water will flow through a slimmer pipe. This principle is basic in considering the choice of a voltage (or pressure) for a distribution system.

There are two general ways of transmitting electric current-overhead and underground. In both cases, the conductor may be copper or

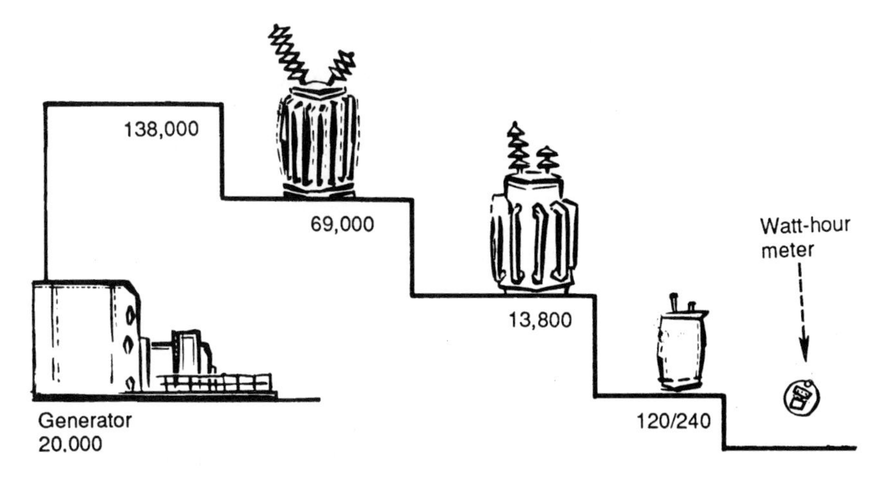

Figure 1-6. Changes in voltage from generator to consumer. All along the journey, voltage is helped down from transmission-line level to a usable level by transformers.

aluminum, but the insulation in the first instance is usually air, except at the supports (poles or towers) where it may be porcelain or glass. In underground transmission, the conductor is usually insulated with rubber, paper, oil, plastic, or other material.

In overhead construction, the cost of the copper or aluminum as compared to the insulation is relatively high. Therefore, it is desirable when transmitting large amounts of electric power, to resort to the higher electrical pressures-or voltages, thereby necessitating slimmer, less expensive conductors. Low voltages necessitate heavy conductors which are bulky and expensive to install, as well as intrinsically expensive.

However, there is a limit to how high the voltage may be made and how thin the conductors. In overhead construction there is the problem of support-poles or towers. If a conductor is made too thin, it will not be able to support itself mechanically. Then the cost of additional supports and pole insulators becomes inordinately high. Underground construction faces the same economic limitation. In this case, the expense is insulation. Underground a cable must be thoroughly insulated and sheathed from corrosion. The higher the voltage, the more insulation is necessary, and the bigger the conductor, the more sheathing is necessary (see Figure 1-7).

Determining distribution voltages is a matter which requires careful studies. Experts work out the system three or four different ways. For instance, they figure all the expenses involved in a 4000-volt (4 kV), in a 34,500-volt (34.5 kV), or a 13,000-volt (13 kV) system.

The approximate costs of necessary equipment, insulators, switches, and so on, and their maintenance and operation must be carefully evaluated. The future with its possibilities of increased demand must also be taken into consideration.

Safety is the most important factor. The National Electric Safety Code includes many limitations on a utility company's choice of voltage. Some municipal areas also set up their own standards.

The utility company must weigh many factors before determining a voltage for distribution.

It was mentioned that safety is the most important factor in determining voltages for distributing electricity. Here's why! Consider what happens when a water pipe carrying water at high pressure suddenly bursts (see Figure 1-8). The consequences may be fatal and damage considerable. The same is true of electrical conductors. Safeguarding the life

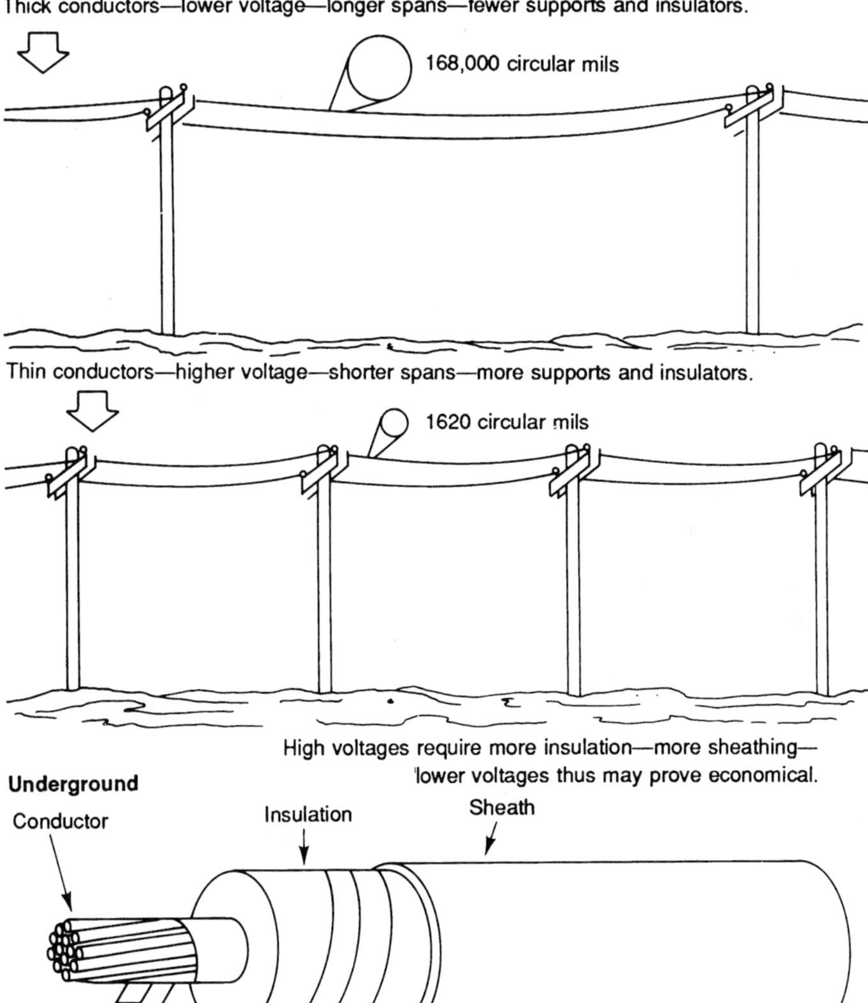

Figure 1-7. Practical economics affect the size of a transmission line.

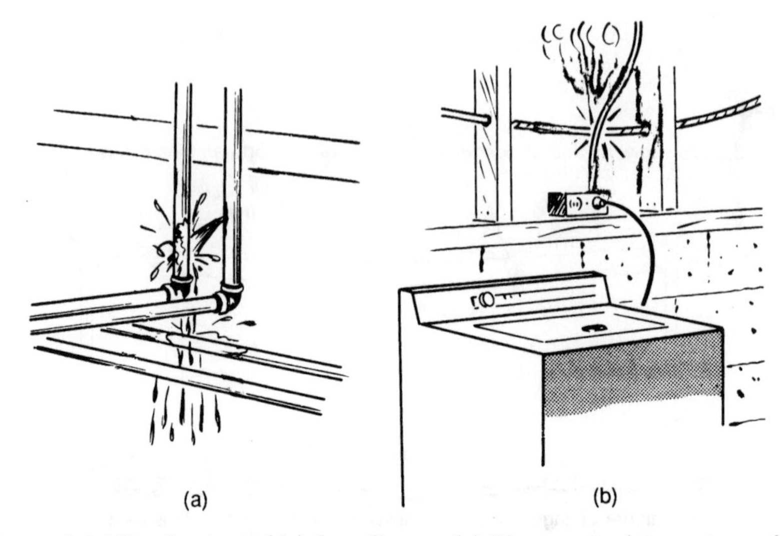

Figure 1-8. The danger of high voltages. (a) Pipe rupturing-water spills over adjacent areas; (b) Rupture of cable insulation causes arcing to other wires-sometimes causing flame.

and limb of the public as well as workers is an important responsibility of the utility company.

Table 1-1 shows typical transmission and distribution system voltages in use at the present time.

Table 1-1. Typical Voltages in Use

Main Transmission	*Sub Transmission*	*Primary Distribution*	*Distribution Secondary*
69,000 V	13,800 V	2,400 V	120 V
138,000 V	23,000 V	4,160 V	120/240 V
220,000 V	34,500 V	13,800 V	240 V
345,000 V	69,000 V	23,000 V	277/480 V
500,000 V	138,000 V	34,500 V	480 V
750,000 V			

REVIEW QUESTIONS

1. What are the three main divisions of an electric power system?

2. Distinguish between transmission and distribution.

3. What is the function of a substation?

4. What are the links between the utility company's facilities and the consumer's premises?

5. What are the two ways of distributing electric energy?

6. Compare the flow of water in a pipe with that of electric current in a wire. What is the relationship between water pressure and voltage? Pipe diameter and conductor diameter? Thickness of pipe and insulation?

7. What is electrical pressure called?

8. What are the two most important factors to be considered in determining a distribution voltage?

9. What changes in the functions of generation and transmission are due to deregulation?

10. What changes in distribution are associated with deregulation?

Chapter 2

Conductor Supports

SUPPORTS FOR OVERHEAD CONSTRUCTION

Conductors need supports to get from one place to another. Supports may be towers, poles, or other structures. The latter may be made of steel, concrete, or wood. The choice of a type of support depends on the terrain to be crossed and the size of conductors and equipment to be carried. Availability and economy, as well as atmospheric elements determine the choice of material.

Usually steel poles and towers (Figure 2-1) are used for transmission lines and wood (Figure 2-2) and concrete poles for distribution circuits. However, this distinction doesn't always hold true. To meet the needs of a particular circumstance, wood or concrete poles can be used to carry transmission lines; and in some instances a steel tower might be necessary for a distribution circuit.

In general, steel towers are used where exceptional strength and reliability are required. Given proper care, a steel tower is good indefinitely.

Steel can also be used for poles. Although they are comparatively expensive, considerations of strength for large spans, crossing railroads or rivers, for example, make wood undesirable and steel poles, complete with steel crossarms are necessary.

In the United States, overhead construction more often consists of wood poles with the conductor wires attached to insulators supported on wood crossarms. Although steel and concrete poles are also used, wood has two desirable advantages: initial economy and natural insulating qualities.

The choice of wood for poles depends on what is available in the particular section of the country. For example, in the central United States, poles of northern white cedar are most apt to be found because it is easily available in Minnesota, Wisconsin, and Michigan. Because of the preponderance of western red cedar in Washington, Oregon, and

Figure 2-1. Steel towers.

Idaho, poles of this wood are found on the Pacific coast. Besides cedar, poles are also made of chestnut or yellow pine. The latter type predominates in the south and east.

TYPES OF POLES

Cedar is one of the most durable woods in this country. It is light, strong, has a gradual taper and is fairly straight, although full of small knots. Before pine poles came into prominence, cedar poles were used where fine appearance in line construction was required. Chestnut is an extremely strong, durable wood and it is not quite as full of knots as cedar, however, chestnut tends to be crooked. The popularity of both cedar and chestnut in past years was largely attributable to their slow rate of decay, particularly at the ground line. The continuous presence of moisture and air, and the chemicals in the soil tend to encourage mouldy growths which consume the soft inner fibers of wood. This is partially offset by treating the butt (that portion of the pole buried in the ground) with a preservative. Although no longer installed, many are in use and will continue to be used for a relatively long time.

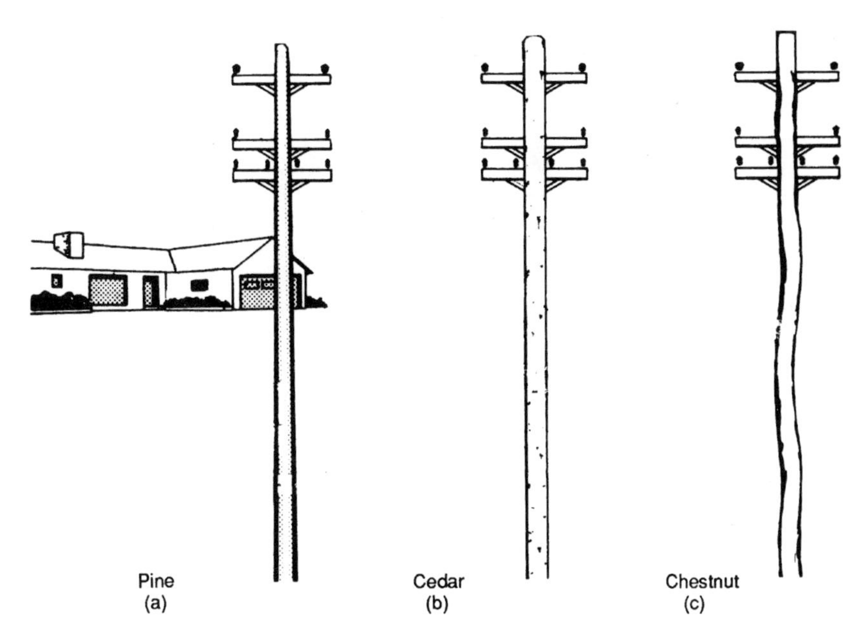

Figure 2-2. Wood poles. (a) Pine: straight, strong, gradual taper—knot free—tendency to decay, offset by treating with preservative, which gives it shiny appearance—prominent east of the Mississippi. (b) Cedar: straight, strong, durable—many small knots—fine appearance—slow decay—prominent in West. (c) Chestnut: strong, durable, but crooked-fewer knots—slow decay—formerly prominent east of the Mississippi.

Long leaf southern yellow pine is very strong, straight, has a gradual taper and is usually fairly free of knots. Despite its excellent appearance, the use of pine in the past was limited by its lack of durability. However, with improvements in wood-preserving methods, yellow pine has come to be widely used. Laminated poles for greater strength are presently being used, particularly for transmission lines, although relatively expensive. Wood pole dimensions are given in Table 2-1.

Metal poles, towers, and structures are subject to rust and corrosion and, hence, must be maintained (painted, parts renewed) periodically. Wood poles and structures decay and are affected by birds and insects, lineworker's climbers, weather, and so on, all of which tend to affect their strength and appearance. To combat decay, poles are in-

Table 2-1. Wood Standard Pole Dimensions

	Class		1	2	3	4	5	6	7
	Min. top circumference		27"	25"	23"	21"	19"	17"	15"
	Min. top diameter		8.6"	8.0"	7.3"	6.7"	6. 1"	5.4"	4.8"
Length of Pole	*Distance Ground Line to Butt*	*Kind of Wood*	*Minimum Circumference at Ground Line (Approximate)*						
25	5	P	34.5	32.5	30.0	28.0	26.0	24.0	22.0
		Ch	37.0	34.5	32.5	30.0	28.0	25.5	24.0
		Wc	38.0	35.5	33.0	30.5	28.5	26.0	24.5
30	5-1/2	P	37.5	35.0	32.5	30.0	28.0	26.0	24.0
		Ch	40.0	37.5	35.0	32.5	30.0	28.0	26.0
		WC	41.0	38.5	35.5	33.0	30.5	28.5	26.5
35	6	P	40.0	37.5	35.0	32.0	30.0	27.5	25.5
		Ch	42.5	40.0	37.5	34.5	32.0	30.0	27.5
		WC	43.5	41.0	38.0	35.5	32.5	30.5	28.0
40	6	P	42.0	39.5	37.0	34.0	31.5	29.0	27.0
		Ch	45.0	42.5	39.5	36.5	34.0	31.5	29.5
		Wc	46.0	43.5	40.5	37.5	34.5	32.0	30.0
45	6-1/2	P	44.0	41.5	38.5	36.0	33.0	30.5	28.5
		Ch	47.5	44.5	41.5	38.5	36.0	33.0	31.0
		WC	48.5	45.5	42.5	39.5	36.5	33.5	31.5
50	7	P	46.0	43.0	40.0	37.5	34.5	32.0	29.5
		Ch	49.5	46.5	43.5	40.0	37.5	34.5	32.0
		WC	50.5	47.5	44.5	41.0	38.0	35.0	32.5
55	7-1/2	P	47.5	44.5	41.5	39.0	36.0	33.5	
		Ch	51.5	48.5	45.0	42.0	39.0	36.0	
		WC	52.5	49.5	46.0	42.5	39.5	36.5	
60	8	P	49.5	46.0	43.0	40.0	37.0	34.5	
		Ch	53.5	50.0	46.5	43.0	40.0	37.5	
		Wc	54.5	51.0	47.5	44.0	41.0	38.5	
65	8-1/2	P	51.0	47.5	44.5	41.5	38.5		
		Ch	55.0	51.5	48.0	45.0	42.0		
		Wc	56.0	52.5	49.0	45.5	42.5		
70	9	P	52.5	49.0	46.0	42.5	39.5		
		Ch	56.5	53.0	48.5	45.5	43.5		
		Wc	57.5	54.0	50.5	47.0	45.0		
75	9-1/2	P	54.0	50.5	47.0	44.0			
		Ch	59.0	54.0	50.0	47.0			
		Wc	59.5	55.5	52.0	48.5			

P—Long Leaf Yellow Pine
Ch—Chestnut Wc—Western Cedar

spected frequently and treated with preservatives.

The question of appearance of such poles and structures is receiving more and more attention. In addition to "streamlining" such installations (as will be discussed later), color has been introduced. For example, wood poles which were formerly brown or black (the "black beauties," oozing creosote), may now be found in green, light blue, tan, or gray colors.

Reinforced concrete poles have become more popular in the United States in areas where concrete proves more economical. (Design drawings of round and concrete poles are shown in Figures 2-3a and 2-3b and Tables 2-2a and 2-2b list their dimensions and strengths.) Usually, the concrete is reinforced with steel, although iron mesh and aluminum can also be used.

POLE LENGTH

Two factors must be considered in choosing poles: length and strength required. The length of poles depends on the required clearance above the surface of the ground, the number of crossarms to be attached, and other equipment which may be installed (Figure 2-4). Provision should also be made for future additions of crossarms, transformers, or other devices. Poles come in standard lengths ranging from 25 to 90 feet in 5-foot differences; that is, 25 feet, 30 feet, 35 feet, and so on. Special poles above 90 feet and below 25 feet are also available.

POLE STRENGTH

Required pole strength is determined by the weight of crossarms, insulators, wires, transformers, and other equipment it must carry, as well as by ice and wind loadings. All these forces tend to break a pole at the ground line.

Ice forms about the conductors and other equipment during a snow or sleet storm (see Figure 2-5). The weight of ice is 57.5 pounds per cubic foot; thus, if ice about 1 inch thick forms about a conductor 100 feet long, more than 100 pounds will be added to the weight carried by the poles. While these direct weights may be appreciable, normal wood poles are more than capable of meeting the ordinary load challenge.

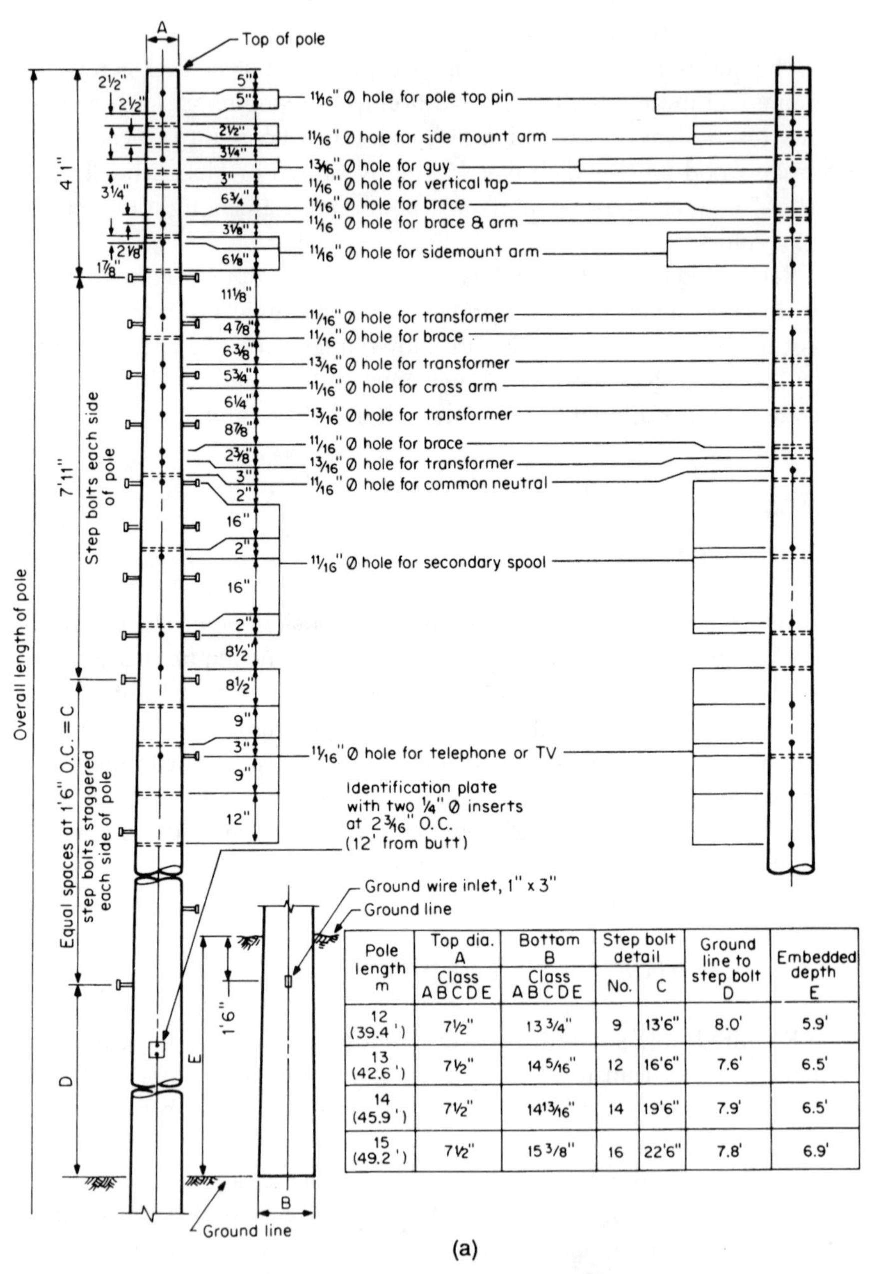

Pole length m	Top dia. A Class ABCDE	Bottom B Class ABCDE	Step bolt detail No.	C	Ground line to step bolt D	Embedded depth E
12 (39.4')	7½"	13 3/4"	9	13'6"	8.0'	5.9'
13 (42.6')	7½"	14 5/16"	12	16'6"	7.6'	6.5'
14 (45.9')	7½"	14 13/16"	14	19'6"	7.9'	6.5'
15 (49.2')	7½"	15 3/8"	16	22'6"	7.8'	6.9'

Figure 2-3(a). Reinforced concrete round hollow distribution pole. Pole steps optional. (c) Use of top full upper part of pole. (*Courtesy Centrecon, Inc.*)

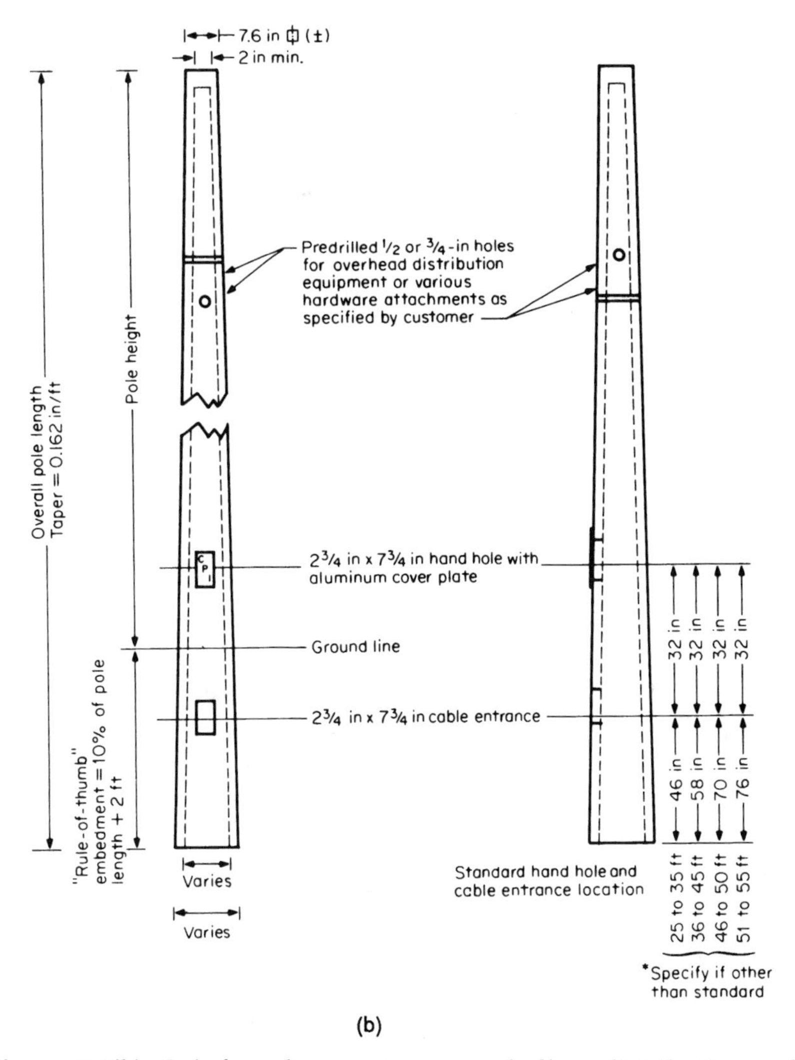

Figure 2-3(b). Reinforced concrete square hollow distribution pole. (*Courtesy Concrete Products, Inc.*)

However, the ice formation about the conductors presents quite a surface to the wind. For example, a 60-mile-per-hour wind, blowing against the ice-coated wire mentioned, will result in a force of more than 135 pounds per conductor being applied to the top of the pole. If this pole suspended three conductors, the total force would be nearly 400 pounds.

Table 2-2a. Dimensions and Strengths—Round Hollow Concrete Poles

Overall length		Class	Setting depth, ft-in	Top diameter, in	Butt diameter, in	Design ultimate moment, ft·lb	Allowable moment—SF = 2, ft·lb	Nominal weight, lb
m	ft-in							
12	39-4	A	5-11	$7\frac{1}{2}$	$13\frac{13}{16}$	100,530	50,260	2540
13	42-8	A	6-3	$7\frac{1}{2}$	$14\frac{5}{16}$	110,130	55,060	2830
14	45-11	A	6-7	$7\frac{1}{2}$	$14\frac{13}{16}$	119,470	59,730	3130
15	49-2	A	6-11	$7\frac{1}{2}$	$15\frac{3}{8}$	128,800	64,400	3470
16	52-6	A	7-3	$7\frac{1}{2}$	$15\frac{7}{8}$	138,400	69,200	3800
12	39-4	B	5-11	$7\frac{1}{2}$	$13\frac{13}{16}$	84,830	42,410	2480
13	42-8	B	6-3	$7\frac{1}{2}$	$14\frac{5}{16}$	92,930	46,460	2760
14	45-11	B	6-7	$7\frac{1}{2}$	$14\frac{13}{16}$	100,800	50,400	3060
15	49-2	B	6-11	$7\frac{1}{2}$	$15\frac{3}{8}$	108,680	54,340	3380
16	52-6	B	7-3	$7\frac{1}{2}$	$15\frac{7}{8}$	116,780	58,390	3710
9	29-6	C	4-11	$6\frac{11}{16}$	$11\frac{7}{16}$	50,810	25,400	1520
10	32-10	C	5-3	$6\frac{11}{16}$	$11\frac{15}{16}$	57,560	28,780	1770
11	36-1	C	5-7	$6\frac{11}{16}$	$12\frac{7}{16}$	64,130	32,060	2010
12	39-4	C	5-11	$7\frac{1}{2}$	$13\frac{13}{16}$	70,690	35,340	2430
13	42-8	C	6-3	$7\frac{1}{2}$	$14\frac{5}{16}$	77,440	38,720	2710
14	45-11	C	6-7	$7\frac{1}{2}$	$14\frac{13}{16}$	84,000	42,000	3000
15	49-2	C	6-11	$7\frac{1}{2}$	$15\frac{3}{8}$	90,560	45,280	3310
16	52-6	C	7-3	$7\frac{1}{2}$	$15\frac{7}{8}$	97,310	48,650	3640
9	29-6	D	4-11	$6\frac{11}{16}$	$11\frac{7}{16}$	41,780	20,890	1490
10	32-10	D	5-3	$6\frac{11}{16}$	$11\frac{15}{16}$	47,330	23,660	1720
11	36-1	D	5-7	$6\frac{11}{16}$	$12\frac{7}{16}$	52,730	26,360	1960
12	39-4	D	5-11	$7\frac{1}{2}$	$13\frac{13}{16}$	58,120	29,060	2390
13	42-8	D	6-3	$7\frac{1}{2}$	$14\frac{5}{16}$	63,670	31,840	2670
14	45-11	D	6-7	$7\frac{1}{2}$	$14\frac{13}{16}$	69,070	34,540	2960
15	49-2	D	6-11	$7\frac{1}{2}$	$15\frac{3}{8}$	74,460	37,230	3260
16	52-6	D	7-3	$7\frac{1}{2}$	$15\frac{7}{8}$	80,010	40,000	3570
9	29-6	E	4-11	$6\frac{11}{16}$	$11\frac{7}{16}$	33,870	16,930	1470
10	32-10	E	5-3	$6\frac{11}{16}$	$11\frac{15}{16}$	38,370	19,180	1700
11	36-1	E	5-7	$6\frac{11}{16}$	$12\frac{7}{16}$	42,750	21,370	1930
12	39-4	E	5-11	$7\frac{1}{2}$	$13\frac{13}{16}$	47,130	23,560	2360
13	42-8	E	6-3	$7\frac{1}{2}$	$14\frac{5}{16}$	51,630	25,810	2640
14	45-11	E	6-7	$7\frac{1}{2}$	$14\frac{13}{16}$	56,000	28,000	2930

(Continued)

Table 2-2a. (*Continued*)

Overall length		Class	Setting depth, ft-in	Top diameter, in	Butt diameter, in	Design ultimate moment, ft·lb	Allowable moment— SF = 2, ft·lb	Nominal weight, lb
m	ft-in							
15	49-2	E	6-11	$7\frac{1}{2}$	$15\frac{3}{8}$	60,380	30,190	3220
16	52-6	E	7-3	$7\frac{1}{2}$	$15\frac{7}{8}$	64,880	32,440	3530
9	29-6	F	4-11	$6\frac{11}{16}$	$11\frac{7}{16}$	27,100	13,550	1460
10	32-10	F	5-3	$6\frac{11}{16}$	$11\frac{15}{16}$	30,700	15,350	1680
11	36-1	F	5-7	$6\frac{11}{16}$	$12\frac{7}{16}$	34,200	17,100	1920
12	39-4	F	5-11	$6\frac{11}{16}$	13	37,700	18,850	2160
13	42-8	F	6-3	$6\frac{11}{16}$	$13\frac{1}{2}$	41,300	20,650	2410
14	45-11	F	6-7	$6\frac{11}{16}$	$14\frac{1}{16}$	44,800	22,400	2680
15	49-2	F	6-11	$6\frac{11}{16}$	$14\frac{9}{16}$	48,300	24,150	2960
9	29-6	G	4-11	$6\frac{11}{16}$	$11\frac{7}{16}$	21,450	10,720	1460
10	32-10	G	5-3	$6\frac{11}{16}$	$11\frac{15}{16}$	24,300	12,150	1670
11	36-1	G	5-7	$6\frac{11}{16}$	$12\frac{7}{16}$	27,080	13,540	1900
12	39-4	G	5-11	$6\frac{11}{16}$	13	29,850	14,920	2150
13	42-8	G	6-3	$6\frac{11}{16}$	$13\frac{1}{2}$	32,700	16,350	2400
14	45-11	G	6-7	$6\frac{11}{16}$	$14\frac{1}{16}$	35,470	17,730	2660
15	49-2	G	6-11	$6\frac{11}{16}$	$14\frac{9}{16}$	38,240	19,120	2930

Courtesy Centrecon. Inc.

It also makes a big difference where the conductor is attached on the pole (see Figure 2-6). A simple illustration of this principle of physics can be seen in many backyards. If wash is hung on a clothesline attached to the top of a pliable pole, it is not surprising to see the pole bend. To keep the pole from bending, the line is attached farther down on the pole.

The same principle applies to poles which must withstand the strain of wind and ice-laden conductors. The higher above the ground the load is applied, the greater will be the tendency for the pole to break at the ground line.

The forces exerted on a line because of ice and wind will depend on climatic conditions, which vary in different parts of the country (see Figure 2-7). In order to safeguard the public welfare, there are published construction standards called the National Electric Safety Code, which

Table 2-2b. Dimensions and Strengths-Square Hollow Concrete Poles

Overall pole length, ft	Pole size-tip/butt, in	E.P.A., ft^2 Concrete strength lb/in^2 6000, standard	7000, when specified	Ultimate ground-line moment, ft - lb	Breaking strength load 2 ft • below tip, lb	Deflection per 100 ft - lb, in	Deflection limitations, ft - lb	Pole weight, lb
25	7.6/11.65	43.3	45.3	84,000	4540	0.03	4100	2260
30	7.6/12.46	38.7	40.7	132,000	5740	0.03	5000	2880
35	7.6/13.27	33.8	36.2	147,000	5350	0.03	5900	3600
40	7.6/14.08	29.4	31.7	163,000	5090	0.03	6800	4370
45	7.6/14.89	25.7	28.1	178,000	4880	0.06	3850	5225
50	7.6/15.70	22.3	24.7	193,000	4710	0.06	4300	6160
55	7.6/16.51	19.1	21.7	209,000	4590	0.06	4750	7270

Glossary of Terms

E.RA. Effective projected area, in square feet of transformers, capacitors, streetlight fixtures, and other permanently attached items which are subject to wind loads. *Concrete strength This is* a reference to the compressive strength of the concrete in pounds per square inch as measured by testing representative samples 28 days after casting.

Ultimate ground-line bending moment This is the bending moment applied to the pole which will cause structural failure of the pole. This is the result of multiplying the load indicated in the column Breaking Strength by a distance 2 feet less than the pole height (i.e., 2 ft. less than the length of pole above ground). Figures under Ultimate Ground-Line Moment assume embedment of 10 percent of the pole length plus 2 feet. The figures in this column on technical charts are maximum moments expected to be applied to the pole. Appropriate safety factors should be used by the designer.

Breaking strength This is the approximate load which, when applied at a point 2 feet below the tip of the pole, will cause structural failure of the pole. *Ground line* The point at which an embedded pole enters the ground or is otherwise restrained.

Deflection The variation at the tip of the pole from a vertical line resulting from the application of loads such as equipment, wind, ice, etc. *Ground-line bending moment* The product of any load applied at any point on the pole multiplied by its height above ground line.

Dead loads This refers to the load on a pole resulting from the attachment of transformers and other equipment permanently.

Live loads These are loads applied to the pole as a result of wind, ice, or other loads of a temporary nature.

(Courtesy Concrete Products Inc.)

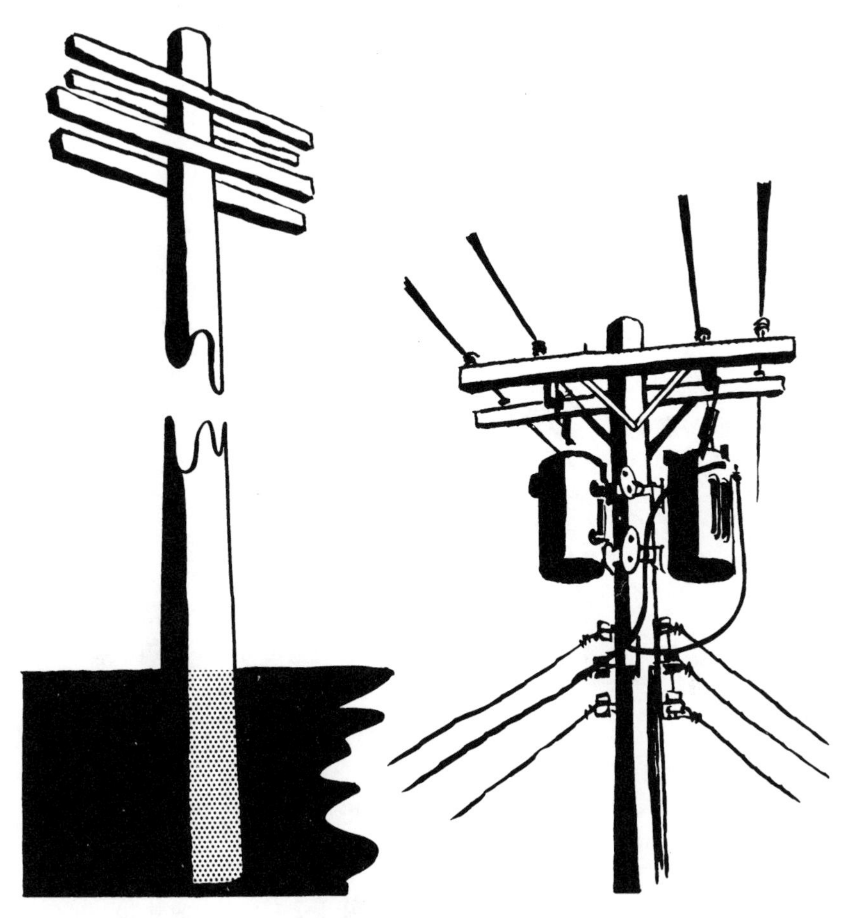

Figure 2-4. Height of poles. Height is dependent on: the number of crossarms, clearance required above ground, and other equipment to be attached.

divides the United States into three loading districts: heavy, medium, and light.

In the heavy loading district, designs of pole lines are based on conductors having a layer of ice (0.5) inch thick, that is, presenting a surface to the wind of the thickness of the conductor plus 1 inch of ice. Wind pressure is calculated at 4 pounds per square foot (that of a 60-mile-per-hour wind) and tension on the conductors is calculated at a temperature of 0°F (–17.8°C). In the medium loading district, these values are reduced to a quarter (0.25) inch of ice and a temperature of 15°F

Figure 2-5. Ice-laden conductors.

(–9.4°C). Wind pressure is calculated at the same 4 pounds per square foot. In the light loading district, no ice is considered, but a wind pressure of 9 pounds per square foot (that of a wind approximately 67-1/2 miles per hour)—and a temperature of 30°F (–2°C) are used for design purposes.

Note that these standards are minimum. As an extra precaution, some companies in the heavy and medium districts calculate with a wind pressure of 8 pounds per square foot and some in the light district use 12 pounds per square foot.

Another factor which contributes to this bending tendency is the force applied by the wires at the poles. Normally, equal spans of wires are suspended from both sides of a pole. However, should the wire span on one side break, or should there be more wire in the span on one side than on the other as shown in Figure 2-8, then the uneven pulls will tend to pull the pole over, again giving the pole a tendency to break at the ground line. These uneven pulls are counteracted by guys which will be discussed later.

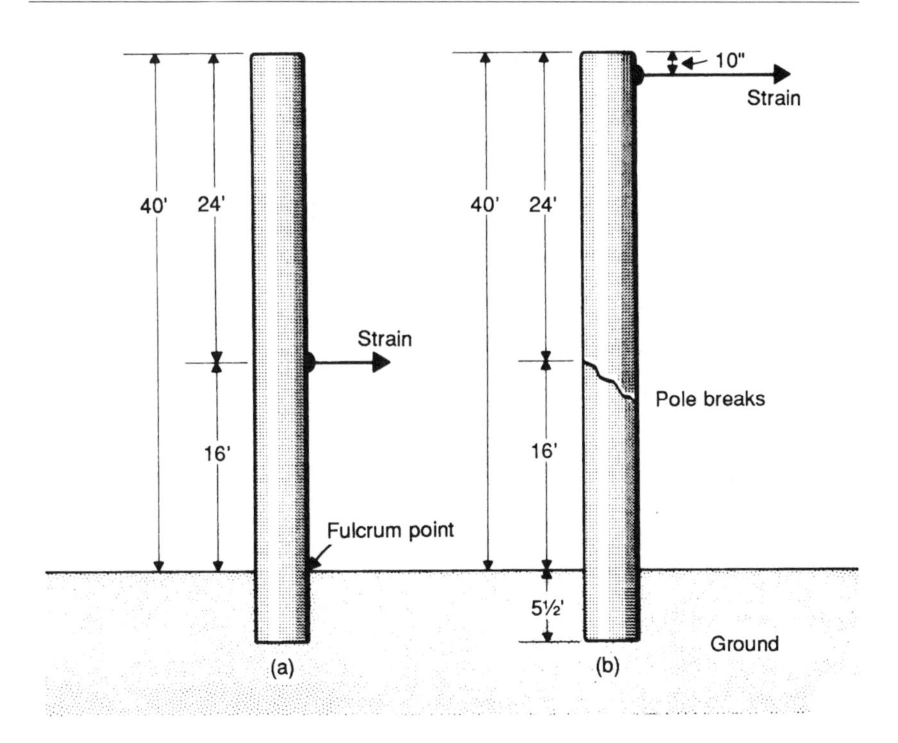

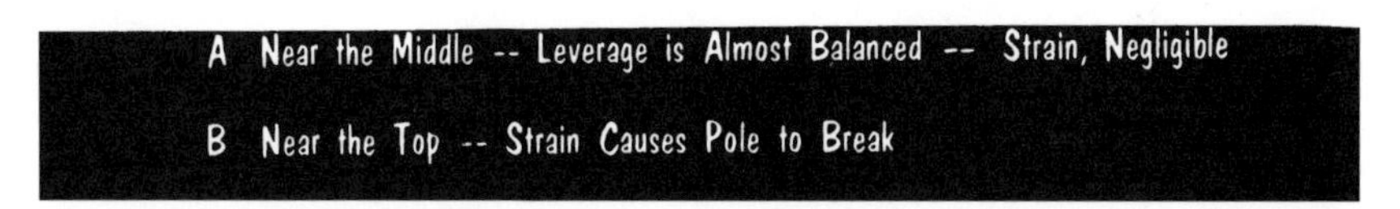

Figure 2-6. Effects of applying loads at different points on a pole. (a) Near the middle, leverage is almost balanced and strain is negligible. (b) Near the top, strain causes the pole to break.

It can be seen that although poles may be the same length, they may have different thicknesses at the ground line to give them varying strengths. However, it is not enough for a pole to be thick enough at the ground line. If it tapers too rapidly, becoming too thin at the top, then the pole may break at some other point. Therefore, in rating poles for strength, a minimum thickness or circumference is specified not only at the ground line, but also at the top. The strength of a pole is expressed as its class. For wood, these classes are usually numbered from 1 to 10 inclusive, class 1 being the strongest. Some extra heavy poles may be of

	Heavy	Medium	Light
Radial thickness of ice (inch)	0.5	0.25	0
Horizontal wind pressure in lb/sq ft	4.	4.	9
Temperature (°F)	0.	+15	+30

Figure 2-7. Wind and ice load specifications. Map shows territorial division of the United States with respect to wind and ice loading of overhead lines. Alaska is in the heavy zone and Hawaii is in the light. ***(National Electric Safety Code)***

class 0, or 00, or even 000. Dimensions are given in Table 2-1 for wood poles and in Tables 2-2a and 2-2b for concrete.

To describe a pole completely, it is necessary to tell the type of material it is made of, its length and its class; for example, 35 feet, class 3, pine pole.

POLE DEPTH

Soil conditions, the height of the pole, weight and pull factors must be considered in deciding how deep a pole must be planted in the ground (Figure 2-9). Table 2-3 gives approximate setting depths for poles in particular given conditions.

For example, suppose a pole 60 feet long is necessary to clear structures or traffic in the path of the conductors. If there are no extra-strain conditions-for example, the ground is solid, the terrain is flat, and the spans are equal-this pole need only be planted 8 feet in the ground.

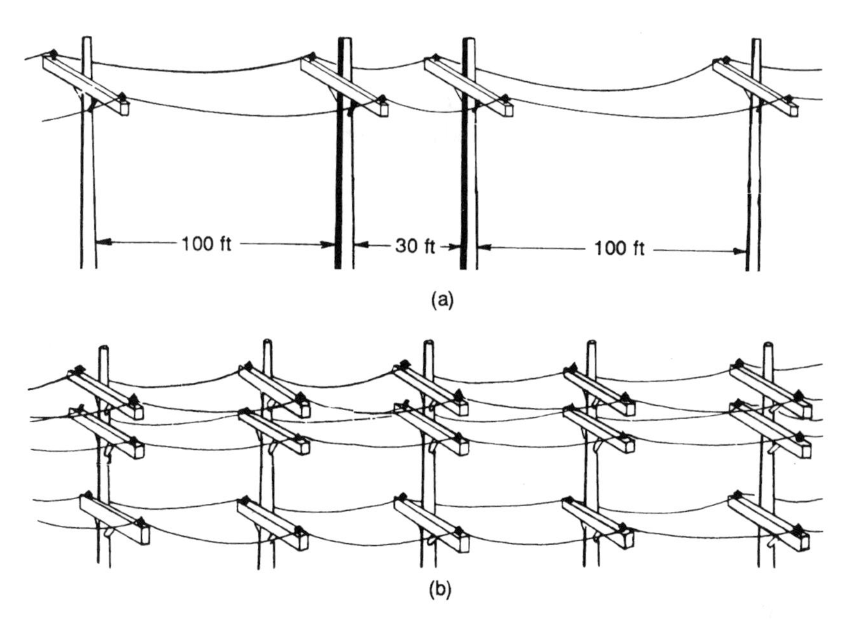

Figure 2-8. Stronger poles counteract bending tendency. (a) With uneven spacing, heavier poles are used to compensate extra pull from longer span. (b) With even spacing, spans balance pulls on each other.

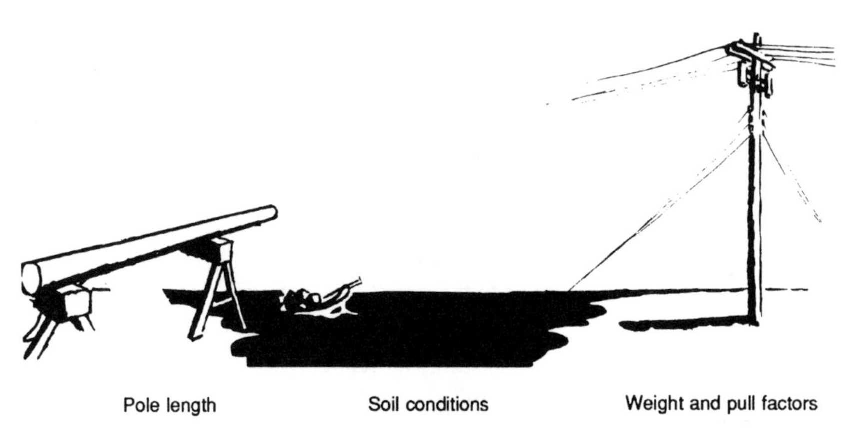

Figure 2-9. Pole depth is determined by pole length, soil conditions, and weight and pull factor.

Table 2-3 Approximate Typical Pole Setting Depths

Length of Pole Overall	*Setting Depths on Straight Lines, Curbs, and Corners*	*Setting Depths at Points of Extra Strain or with Poor Soil Conditions*
30 ft and under	5 ft 0 in.	5 ft 6 in.
35 ft	5 ft 6 in.	6 ft 0 in.
40 ft	6 ft 0 in.	6 ft 6 in.
45 ft	6 ft 6 in.	7 ft 0 in.
50 ft	7 ft 0 in.	7 ft 6 in.
55 ft	7 ft 6 in.	8 ft 0 in.
60 ft	8 ft 0 in.	8 ft 6 in.
65 ft	8 ft 0 in.	8 ft 6 in.
70 ft and over	Special Setting Specified	

However, if there is an unequal span of wire on one side creating a strain or if the soil conditions are poor, the pole must be set 8-1/2 feet deep.

POLE GAINS

Gaining is the process of shaving or cutting a pole to receive the crossarms. In some cases this consists of Cutting a slightly concaved recess 1/2 inch in depth [see Figure 2-10(a)] so that the arm cannot rock.

The recess in the pole is now considered unnecessary. Instead the surface of the pole where the crossarm will be attached is merely flattened to present a flat smooth area. This is called "slab gaining," see Figure 2-10(b). The cross arm is then fastened to the pole with a through bolt. Two flat braces are attached to secure the arm. Some poles have two crossarms mounted one on either side of the pole. These are known as double arms. When double arms are installed, the pole is gained on one side only. A double gain would tend to weaken the pole and is unnecessary since the tightening of the through-bolt causes the arm on the back of the pole to bite into the surface.

Pole steps for the lineworker to climb are usually installed at the same time that the pole is roofed and gained, where these are considered to be desirable.

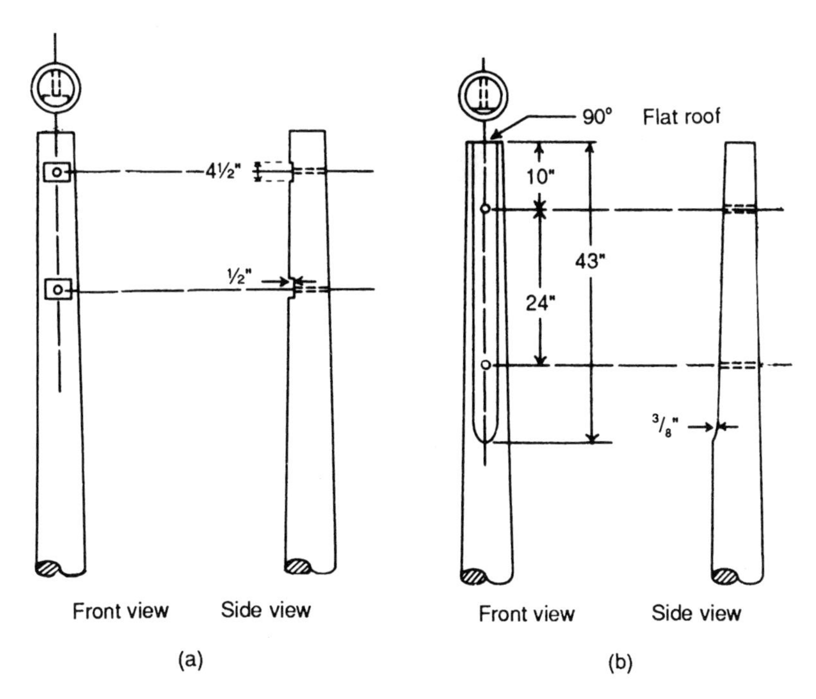

Figure 2-10 (a). Recess-gained pole and (b) slab-gained pole. Measurements are for poles 25 to 60 feet long. For poles 65 feet long and over, slab gain should total 73 inches in length.

CROSSARMS

The woods most commonly used for crossarms are Douglas Fir or Longleaf Southern Pine because of their straight grain and durability. The top surface of the arm is rounded [see Figure 2-11 (a)] so that rain or melting snow and ice will run off easily.

The usual cross-sectional dimensions for distribution crossarms are 3-1/2 inches by 4-1/2 inches; their length depending on the number and spacing of the pins. Heavier arms of varying lengths are used for special purposes, usually for holding the heavier transmission conductors and insulators. Four-pin [see Figure 2-11 (b)], six-pin, and eight-pin arms are standard for distribution crossarms, the six-pin arm being the most common. Where unusually heavy loading is encountered, as at corner or junction poles, double arms, that is, one on each side of the pole may be required as shown on Figure 2-11 (c). Again, the emphasis on appear-

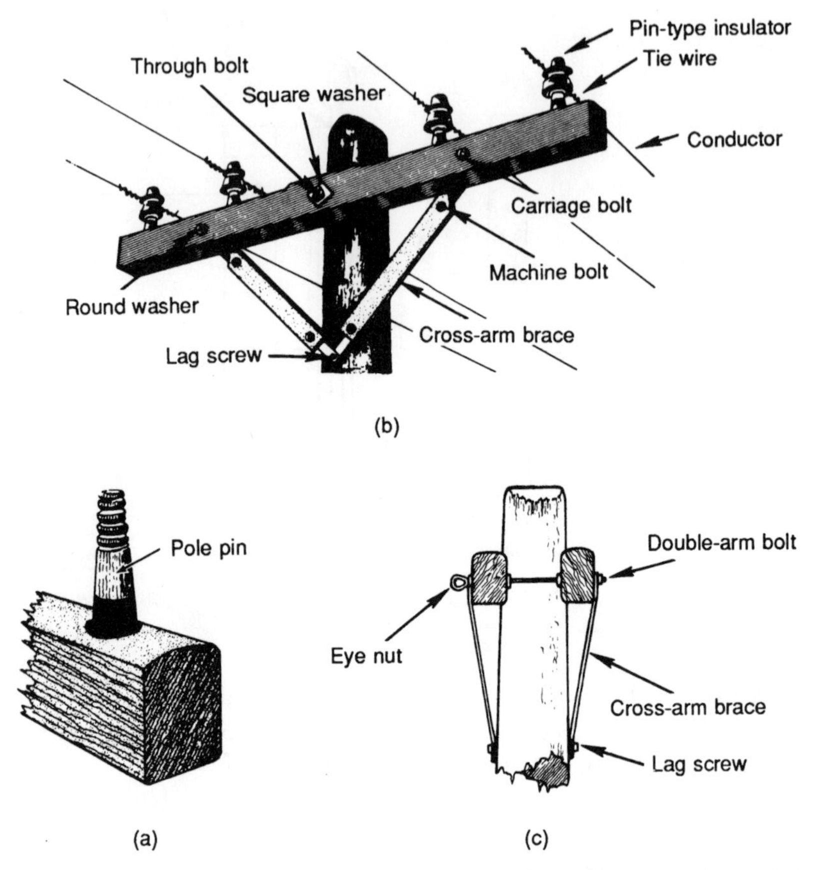

Figure 2-11 (a). Cross-sectional view of rounded off top surface of the arm. (b) A single distribution arm mounting on a pole. (c) Profile cross section showing pole-mounting of a double crossarm.

ance has caused construction designs to eliminate the installation of crossarms, as will be discussed later.

POLE PINS

Pole pins shown in Figure 2-12 are attached to the crossarms. They are used to hold pin-type insulators. Note that they are threaded so that the insulator (discussed in Chapter 3) can be securely screwed on.

Yellow or black locust wood is most commonly used for attaching

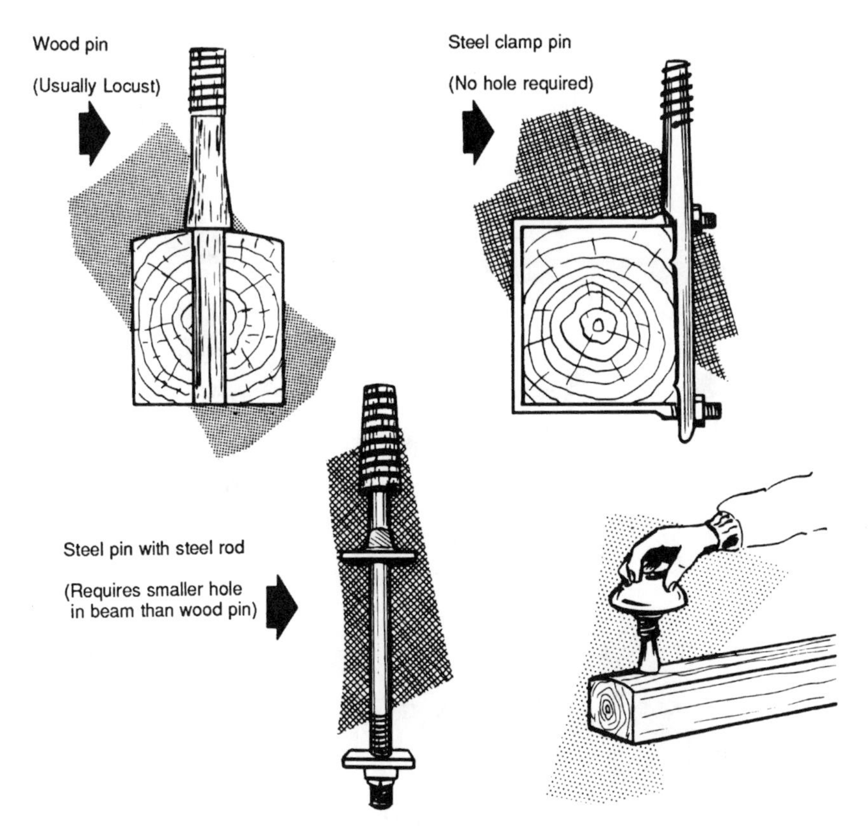

Figure 2-12. Pole pins for attaching pin insulators to crossarms.

the insulators to the crossarms because of its strength and durability; steel pins are used where greater strength is required.

PIN SPACING

The spacing of the pins (see Figure 2-13) on the crossarms must be such as to provide enough air space between the conductors to prevent the electric current from jumping or flashing over from one conductor to another. Also, sufficient spacing is necessary to prevent contact between the wires at locations between poles when the wires sway in the wind. In addition, enough space must be provided to enable workers climbing through the wires to work safely. The spacing on a standard sixpin arm is 14-1/2 inches, with 30 inches between the first pins on either side of

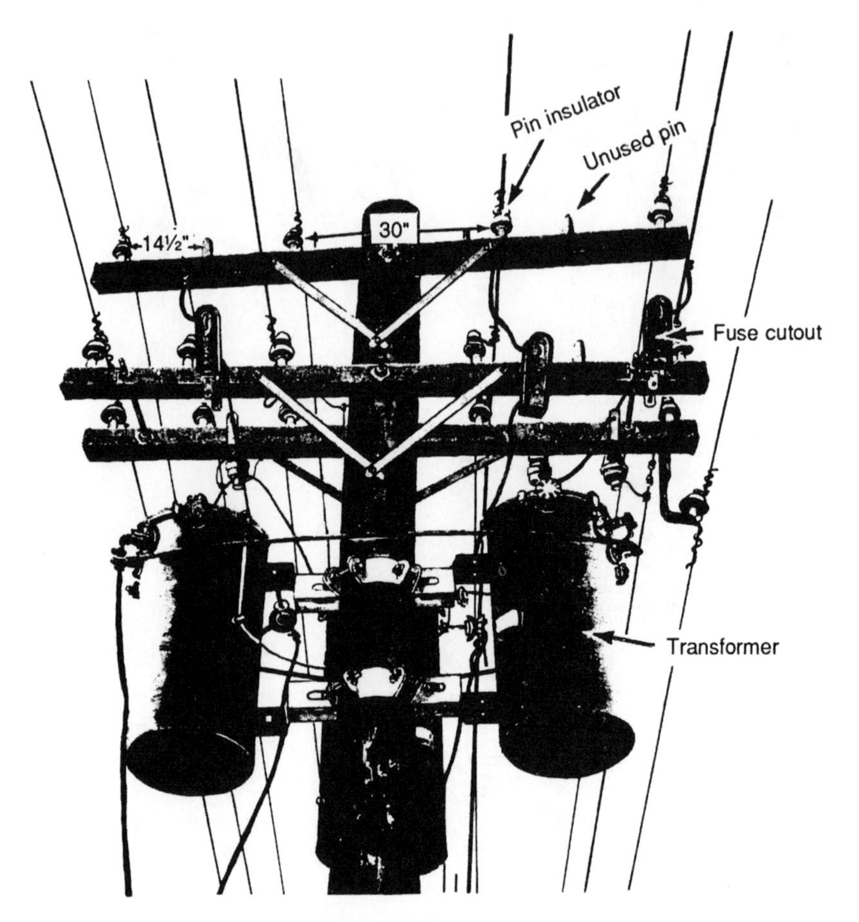

Figure 2-13. Pin spacing shown for six-pin arms mounted on a pole.

the pole for climbing space. A special six-pin arm with spacing wider than 30 inches is frequently used for junction poles to provide greater safety for the workers.

SECONDARY CABLE

Economy and appearance have dictated the use of wires twisted into a cable for use as secondary mains and for service connections to buildings (Figure 2-14). The wires carrying current are generally insulated with a plastic material; the neutral conductor is very often left bare

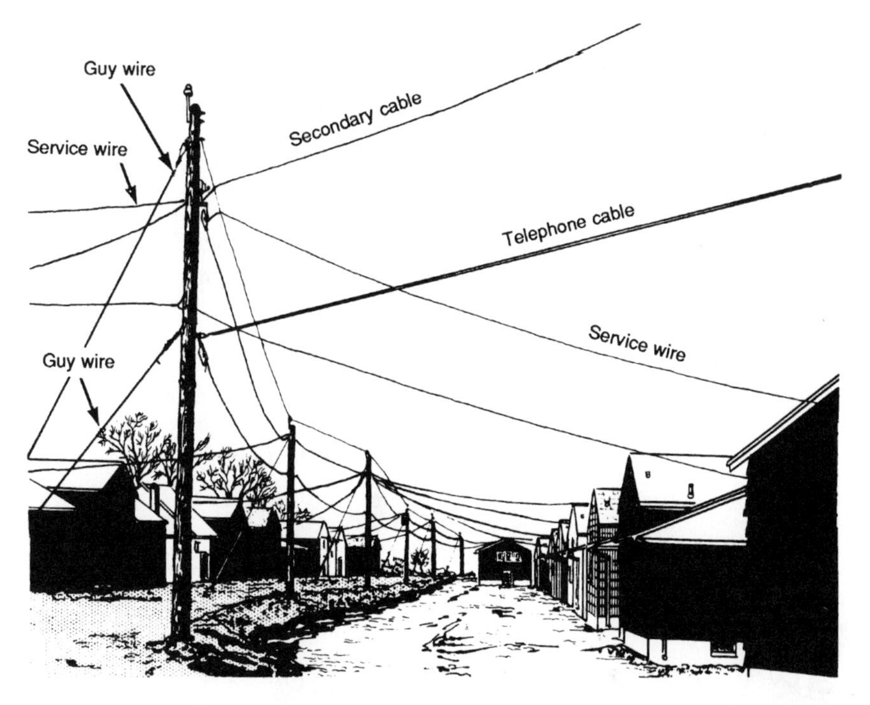

Figure 2-14. Secondary cable installation.

and may act as the supporting wire for the cable. The twisted combination, or bundle, is strung from pole to pole as a secondary main, or from pole to building as a service drop or connection. The bundle may consist of two wires (duplex), three wires (triplex), or four wires (quadruplex).

This cable became economical and feasible when manufacturers developed a means of connecting cable to the house service wires without separating the conductors on the pole, as is necessary with secondary racks. These connectors save man-hours and pole space. To install four services, the utility company need only install a bracket, a neutral connector, and two-phase connectors. With the secondary rack, *each service* required the installation of a clamp, a hook, and three connectors.

SECONDARY RACKS

Secondary mains were often supported in a vertical position. When so supported, a so-called secondary rack was used in place of the crossarm (see Figure 2-15). In this type of support, the conductors are spaced closely and are strung on one side of the pole. As the electrical pressure or voltage between these conductors is relatively low, usually 120 or 240 volts, it is not necessary to maintain the same spacing as on crossarms.

The use of this secondary rack simplifies the installation of service wires to the consumer's premises. When a number of services run from each side of the pole, a second rack is installed on the opposite side for support of the services on that side.

REVIEW QUESTIONS

1. What are the major components of overhead construction?

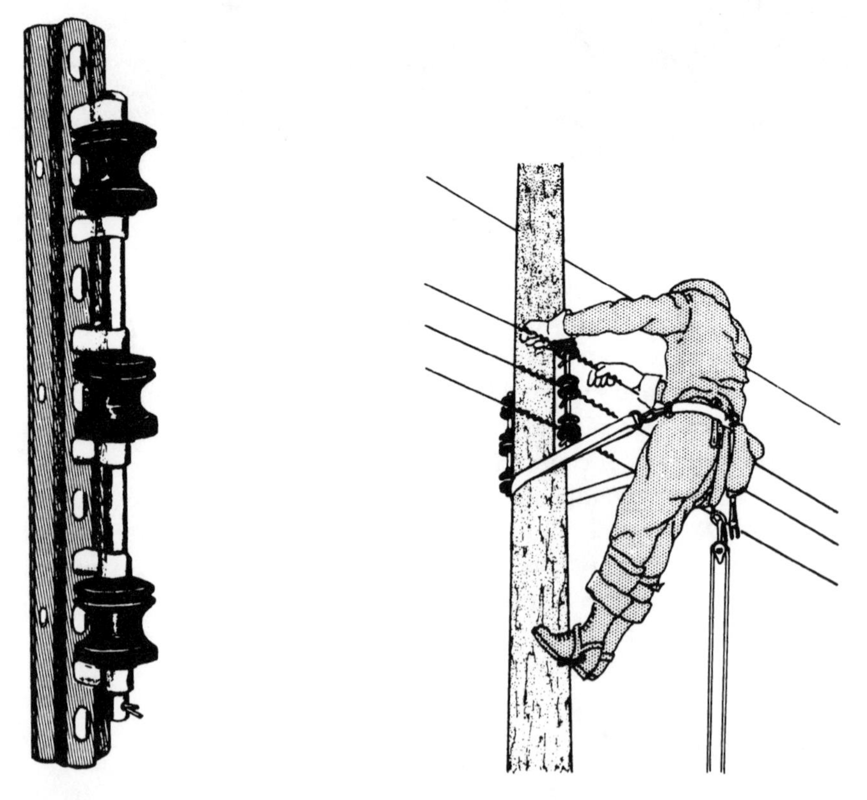

Figure 2-15. A secondary rack holding spool insulators. Lineworker tying wire around insulators on secondary rack.

2. Conductor supports are made of three materials. Name them.
3. Of what kinds of material are poles generally made?
4. How is pole decay reduced to a minimum?
5. What factors should be considered in determining the length of poles to be used in overhead construction?
6. What factors should be considered in determining the strength of the pole required?
7. How is a pole described completely?
8. Define pole *gaining*.
9. What are the functions of crossarms and pole pins?
10. What is a secondary rack? What are the advantages of secondary cable?

Chapter 3

Insulators and Conductors

INSULATORS

Line conductors are electrically insulated from each other as well as from the pole or tower by nonconductors which are called insulators (see Figure 3-1).

To determine whether or not an insulator can be used, both its mechanical strength and electrical properties must be considered. Two practical insulator materials are porcelain and glass. Both of these leave much to be desired. Porcelain can withstand heavy loading in compression, but tears apart easily under tension, that is, when pulled apart. In using a porcelain insulator, therefore, care must be taken to make the forces acting on it compress and not pull apart. The same is generally true of glass.

Although glass insulators are good for lower voltage applications, porcelain insulators are much more widely used because they are more practical (see Figure 3-2). Porcelain has two advantages over glass: (1) it can withstand greater differences in temperature, that is, it will not crack

Figure 3-1. Insulators: (a) post type and (b) pin type.

when subjected to very high or very low temperatures; (2) porcelain is not as brittle as glass and will not break as easily in handling or during installation.

Polymer insulators are not so restricted and have the advantage also of being lighter in weight than porcelain or glass.

Pin-Type Insulators

Insulators, in compression, supporting conductors may be classified as pin type and post type.

The pin-type insulator is designed to be mounted on a pin which in turn is installed on the crossarm of the pole. The insulator is screwed on the pin and the electrical conductor is mounted on the insulator (Figure 3-3). Made of porcelain or glass, the pin insulator can weigh anywhere from 1/2 pound to 90 pounds.

This type of insulator is applicable for rural and urban distribution circuits, and it is usually constructed as one solid piece of porcelain or glass. In Figure 3-4, note the grooves for the conductor and for the tie wires.

Larger, stronger pin-type insulators are used for high-voltage transmission lines. These differ in construction in that they consist of two or three pieces of porcelain cemented together. These pieces form what are called petticoats. They are designed to shed rain and sleet easily.

Post-type insulators are somewhat similar to pin-type insulators. They are generally used for higher voltage applications with the height and number of petticoats being greater for the higher voltages. They may be mounted horizontally as well as vertically, although their strength is diminished when mounted horizontally. The insulator is

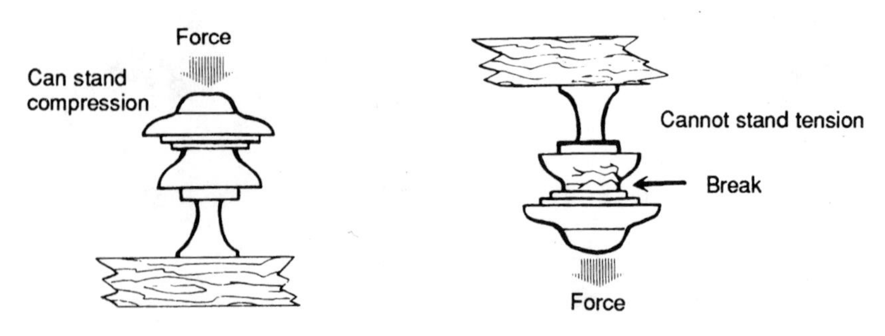

Figure 3-2. How porcelain reacts to compression and tension forces.

Insulators

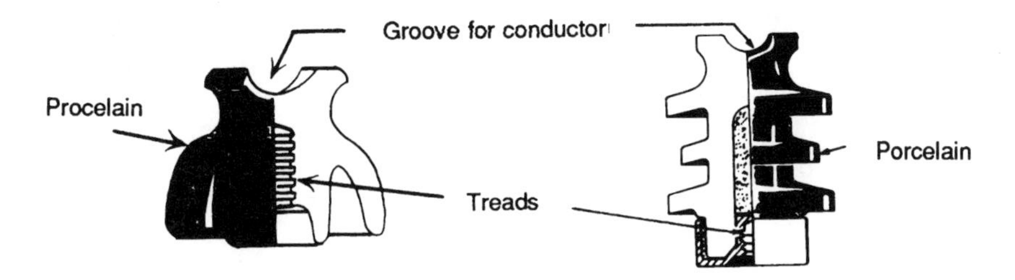

Figure 3-3. Two styles of low-voltage porcelain insulator.

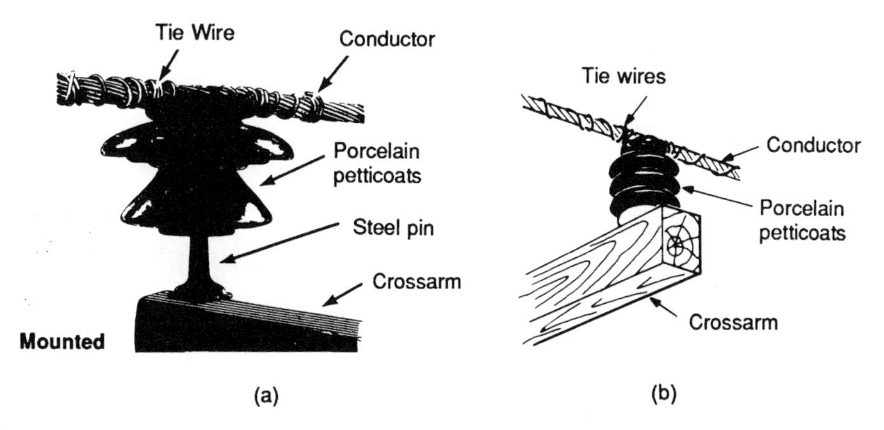

Figure 3-4 (a). Pin and (b) post-type insulators.

made of one piece of porcelain and its mounting bolt or bracket is an integral part of the insulator.

Advantage of Pin or Post over Suspension Insulators

The most commonly used insulators are the pin or post type and the suspension (or hanging) type [see Figure 3-4(c)]. A third type, the strain insulator, is a variation of the suspension insulator and is designed to sustain extraordinary pulls. Another type, the spool insulator, is used with secondary racks and on service fittings.

The main advantage of the pin or post-type insulator is that it is cheaper. Also, the pin or post insulator requires a shorter pole to achieve the same conductor above the crossarm while the suspension insulator suspends it below the crossarm.

Figure 3-4 (c) Advantage of pin and post insulators over suspension insulators. Same conductor height is achieved with shorter pole.

Post-Type Insulators for Armless Construction

Post-type insulators are also used to support conductors of a polyphase primary circuit (and some transmission lines) without use of cross arms, as shown in Figure 3-5. This type of installation is employed where appearance or narrow rights-of-way considerations are important.

Overhead Construction Specifications

Refer to Figures 5-25 through 5-31 at the end of Chapter 5 for typical construction standards of primary line assemblies.

Suspension Insulators

The higher the voltage, the more insulation is needed. Transmission lines use extremely high voltages, 69,000 to 375,000 volts, for example. At these voltages the pin or post-type insulator becomes too bulky and cumbersome to be practical, and the pin which must hold it would have to be inordinately long and large. To meet the problem of insulators for these high voltages, the suspension insulator shown in Figure 3-6 was developed.

The suspension insulator *hangs* from the crossarm, as opposed to the pin insulator which sits on top of it. The line conductor is attached to its lower end. Because there is no pin problem, any distance can be put between the suspension insulator and the conductor can be provided just by adding more insulators to the "string,"

The entire unit of suspension insulators is called a string. How many insulators this string consists of depends on the voltage, the weather conditions, the type of transmission construction, and the size of insulator used. It is important to note that in a string of suspension insulators one or more insulators can be replaced without replacing the whole string.

Strain Insulators

Sometimes a line must withstand great strain as shown in Figure 3-7, for instance at a corner, at a sharp curve, or at a dead-end. In such a circumstance the pull is sustained and insulation is provided by a strain insulator. On a transmission line, this strain insulator often con-

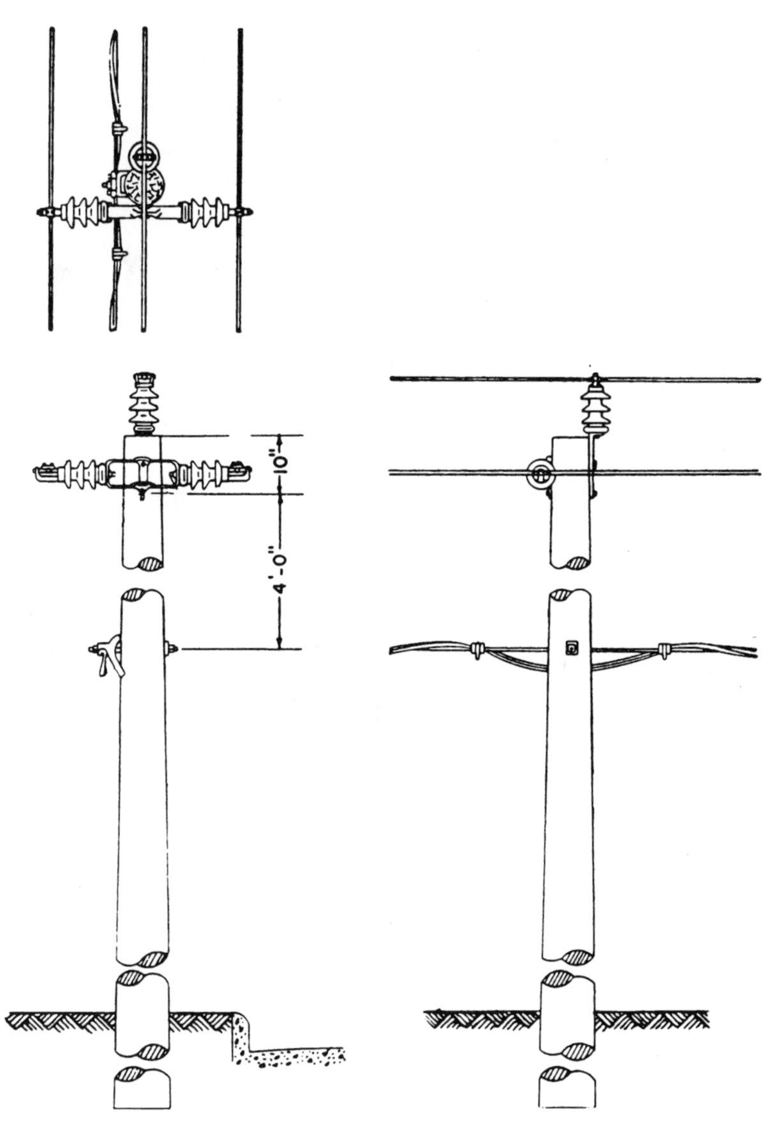

Figure 3-5. 3 phase 13 kV through line pole armless construction 20 kV insulators. *(Courtesy Long Island Lighting Co.)*

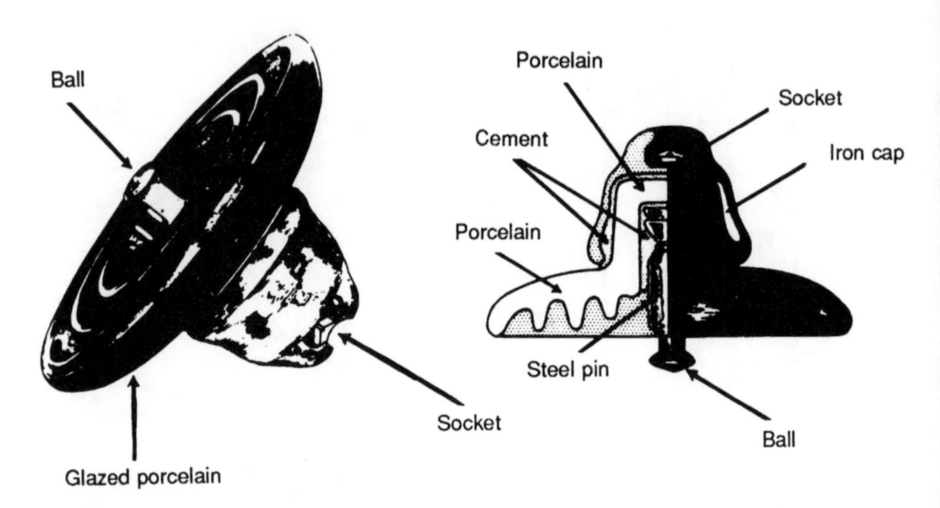

Figure 3-6. Ball and socket-type suspension insulator.

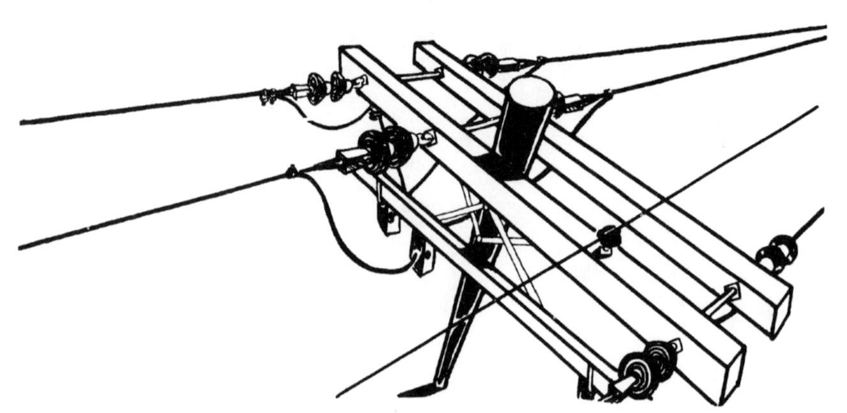

Figure 3-7. How a suspension insulator is used to withstand strain.

sists of an assembly of suspension insulators. Because of its peculiarly important job, a strain insulator must have considerable strength as well as the necessary electrical properties. Although strain insulators come in many different sizes, they all share the same principle; that is, they are constructed so that the cable will compress (and not pull apart) the porcelain.

Strain insulators are sometimes used in guy cables, where it is necessary to insulate the lower part of the guy cable from the pole for the safety of people on the ground. This type usually consists of a porcelain piece pierced with two holes at right angles to each other through which the two ends of the guy wires are looped, in such a manner that the porcelain between them is in compression, Figure 3-8.

Figure 3-8. Strain insulator used for guy wires. Wires of insulator pull in opposite directions, resulting in compression.

Spool Insulators

The spool-type insulator, which is easily identified by its shape, is usually used for secondary mains. The spool insulator may be mounted on a secondary rack or in a service clamp as shown in Figures 3-9(a) and 3-9(b). Both the secondary low-voltage conductors and the house service wires are attached to the spool insulator. The use of such insulators has decreased greatly since the introduction of cabled secondary and service wires.

The tapered hole of the spool insulator distributes the load more evenly and minimizes the possibility of breakage when heavily loaded. The "clevis" which is usually inserted in this hole is a piece of steel metal with a pin or bolt passing through the bottom.

POLYMER (PLASTIC) INSULATION

In general, any form of porcelain insulation may be replaced with polymer. Their electrical characteristics are about the same. Mechanically they are also about the same, except porcelain must be used under compression while polymer may be used both in tension and compression. Porcelain presently has an advantage in shedding rain, snow and ice and in its cleansing effects; polymer may allow dirt, salt and other pollutants to linger on its surface encouraging flashover, something that can be overcome by adding additional polymer surfaces. Polymer's greatest advantage, however, is its comparatively light weight making for labor savings; and essentially no breakage in handling. For detailed comparison of the two insulations, refer to Appendix A, Insulation: Porcelain vs. Polymer.

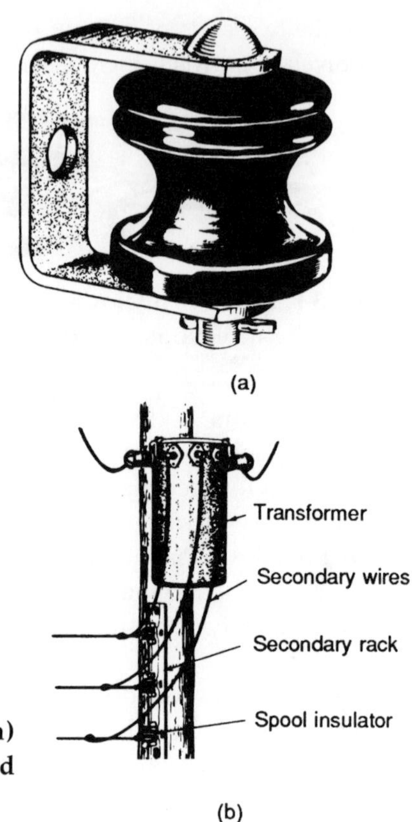

Figure 3-9. Spool insulator (a) mounted in clamp, and (b) mounted on a secondary rack.

CONDUCTORS

Line Conductors

Line conductors may vary in size according to the rated voltage. The number of conductors strung on a pole depends on the type of circuits that are used.

Because they strike a happy combination of conductivity and economy, copper, aluminum, and steel are the most commonly used conductor materials. Silver is a better conductor than copper; but its mechanical weakness and high cost eliminate it as a practical conductor (see Figure 3-10).

On the other hand, there are cheaper metals than copper and aluminum; but they would be hopelessly poor conductors. Copper is the touchstone of conductors. Other conducting materials are compared to

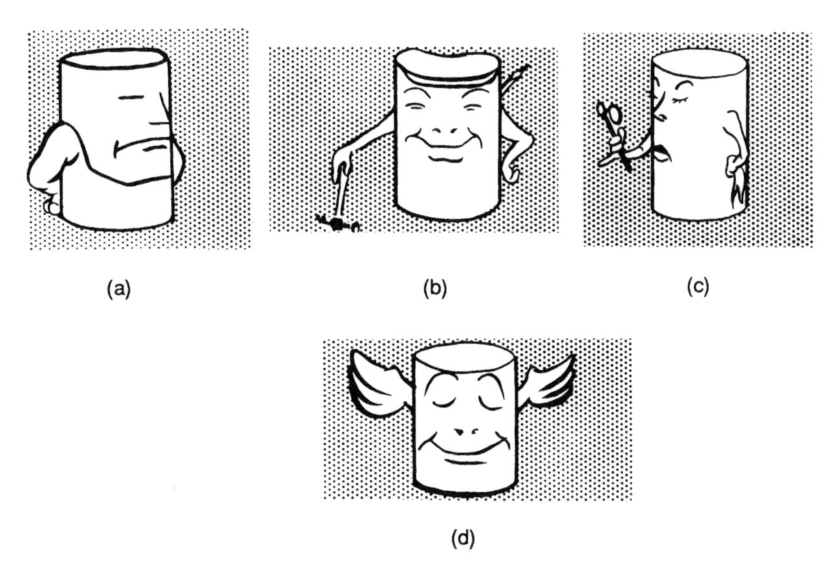

Figure 3-10. All conductors are compared to copper. (a) Steel, strong, but poor conductor. (b) Copper, an all-around good, durable conductor for its price. (c) Aluminum, light, durable, rust-proof, but only 70% as good a conductor. (d) Silver, better conductor, too expensive, and mechanically weak.

copper to determine their economic value as electrical conductors.

Aluminum-steel or copper-steel combinations and aluminum have become popular for conductors in particular circumstances. Aluminum alloys are also used as conductors.

Copper Conductors

Copper is used in three forms: hard drawn, medium-hard drawn, and soft drawn (annealed). Hard-drawn copper wire has the greatest strength of the three and is, therefore, mainly used for transmission circuits of long spans (200 ft or more). However, its springiness and inflexibility make it hard to work with [see Figure 3-11 (a)].

Soft-drawn wire is the weakest of the three [Figure 3-11 (b)]. Its use is limited to short spans and for tying conductors to pin-type insulators. Since it bends easily and is easy to work with, soft-drawn wire is used widely for services to buildings and some distribution circuits. Practice, however, has been toward longer distribution circuit spans and use of medium-hard-drawn copper wire.

Aluminum and ACSR Conductors

Aluminum is used because of its light weight as illustrated in Figure 3-12, which is less than one third that of copper. It is only 60 to 80 percent as good a conductor as copper and only half as strong as copper. For these reasons it is hardly ever used alone, except for short distribution spans. Usually the aluminum wires are stranded on a core of steel wire. Such steel reinforced aluminum wire has great strength for the weight of the conductor and is especially suitable for long spans. Transmission lines often consist of aluminum conductors steel reinforced (ACSR).

Steel Conductors

Steel wire is rarely used alone. However, where very cheap construction is needed, steel offers an economic advantage. Because steel wire is three to five times as strong as copper, it permits longer spans and requires fewer supports. However, steel is only about one tenth as good a conductor as copper and it rusts rapidly [see Figure 3-13(a)]. This rusting tendency can be counteracted (so that steel wire will last longer) by galvanizing, that is, by the application of a coat of zinc to the surface.

Copperweld or Alumoweld Conductors

The main disadvantages of steel are a lack of durability and con-

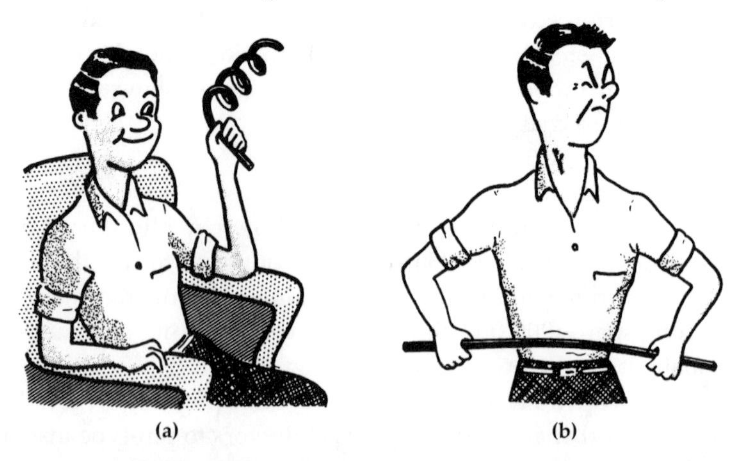

Figure 3-11 (a). Hard-drawn copper is strong but inflexible and springy. It is used for long spans. (b) Soft-drawn (annealed) copper is weak but easy to handle. It is used for service and some distribution circuits.

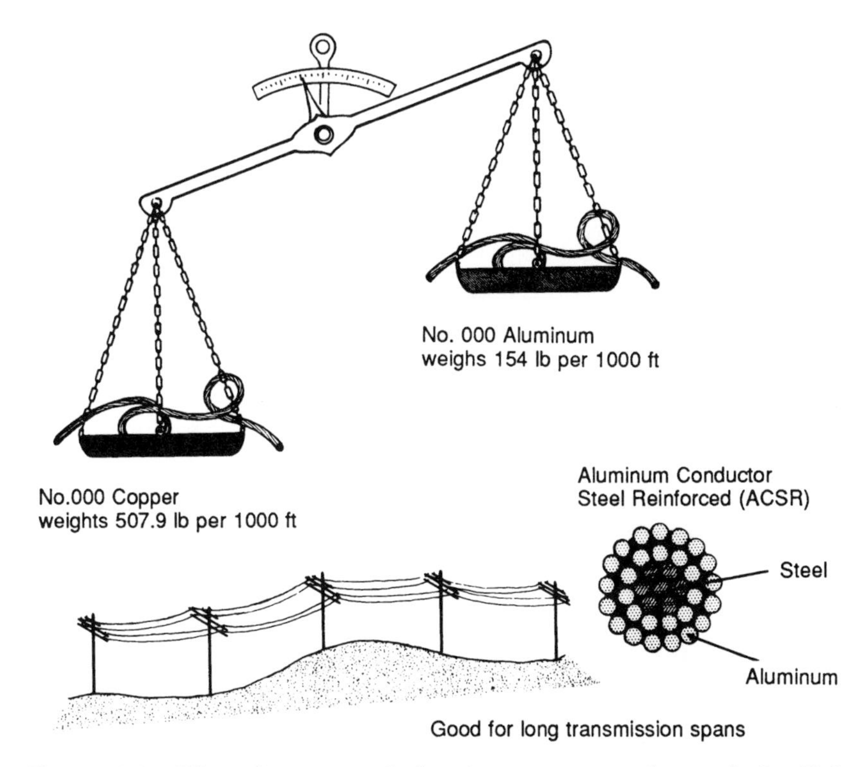

Figure 3-12. The advantage of aluminum as a conductor is its light weight-less than one third the weight of copper.

ductivity. On the other hand, steel is cheap, strong, and available. These advantages made the development of copper-clad or aluminum-clad steel wire [see Figure 3-13(b)] most attractive to the utility companies. To give steel wire the conductivity and durability it needs, a coating of copper is securely applied to its outside. The conductivity of this clad steel wire can be increased by increasing the coating of copper or aluminum. This type of wire, known as Copperweld or Alumoweld is used for guying purposes and as a conductor on rural lines, where lines are long and currents are small.

Conductor Stranding

As conductors become larger, they become too rigid for easy handling. Bending can damage a large solid conductor. For these practical reasons, the stranded conductor was developed (see Figure 3-14). A stranded conductor consists of a group of wires twisted into a single

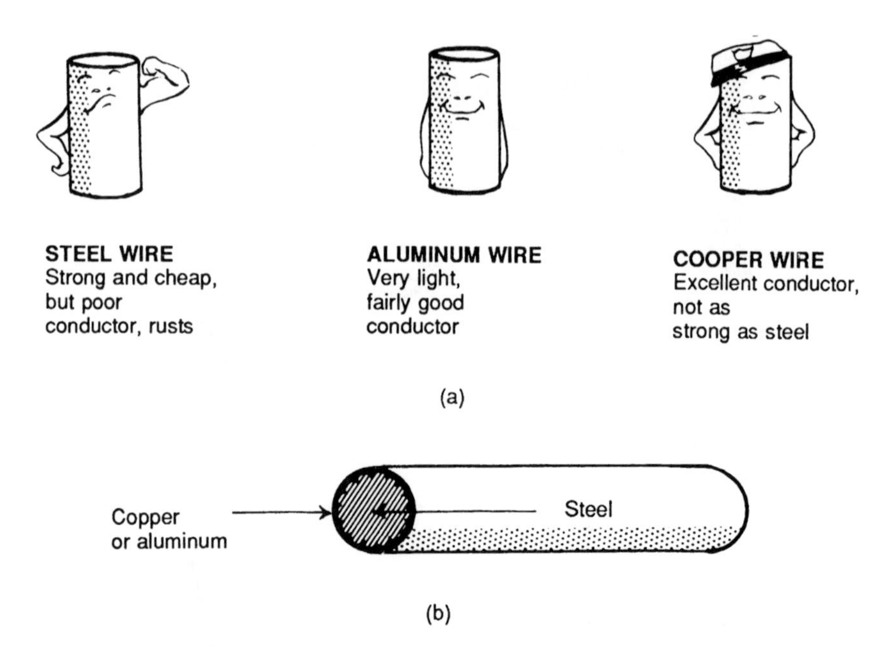

Figure 3-13 (a). Comparison of metal conductors. (b) Copperweld or Alumoweld conductor used for rural distribution and guy wires.

conductor. The more wires in the conductor's cross section, the greater will be its flexibility. Usually, all the strands are of the same size and same material (copper, aluminum, or steel). However, manufacturers do offer stranded conductors combining these metals in different quantities.

Sometimes 3 strands of wire are twisted together. But usually, they are grouped concentrically around 1 central strand in groups of 6. For example, a 7-strand conductor consists of 6 strands twisted around 1 central wire. Then 12 strands are laid over those 6 and twisted to make a 19-strand conductor.

To make a 37-strand conductor, 18 more are placed in the gaps between these 18. And so the number of strands increases to 61, 91, 127, and so on.

Other combinations are possible. For example, 9 strands can be twisted around a 3-strand twist to make a 12-strand wire. Two large strands can be twisted slightly and then surrounded by 12 twisted strands making a 14-strand conductor.

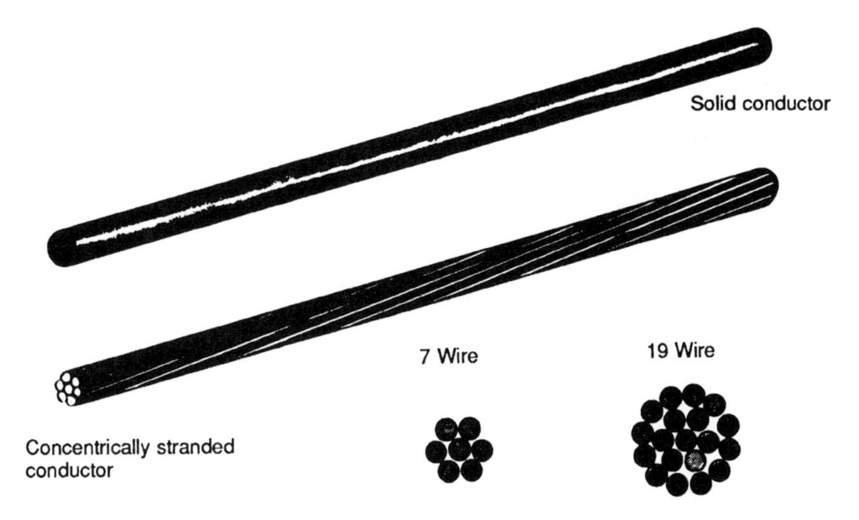

Figure 3-14. Types of conductors. Cross section of stranded conductors.

Conductor Coverings

Conductors on overhead lines may be either bare or covered. Such conductors, located in trees, or adjacent to structures where they may come into occasional contact, may be covered with high density polyethylene or other plastic material resistant to abrasion. This covering is generally not sufficient to withstand the rated voltage at which the line is operating; so these conductors must be mounted on insulators anyway. The purpose of the covering is mainly to protect the wire from mechanical damage. The wires should be treated as though they were bare.

Whether or not the lines are covered, workers consider it necessary for safety to work with rubber. Besides wearing rubber gloves (see Figure 3-15) and sleeves, they make sure to cover the conductors, insulators, and other apparatus with line hose, hoods, blankets, and shields.

Developments now allow workers to work on conductors while energized, as long as the platform or bucket in which they are standing is insulated. In this "bare hand" method, they usually wear noninsulating leather work gloves. Extreme caution is necessary in this method, as the insulated platform or bucket insulates workers from a live conductor and ground, but does not protect them when working on two or more live conductors between which high voltage may exist.

Transmission lines, operating at higher voltages, may be worked

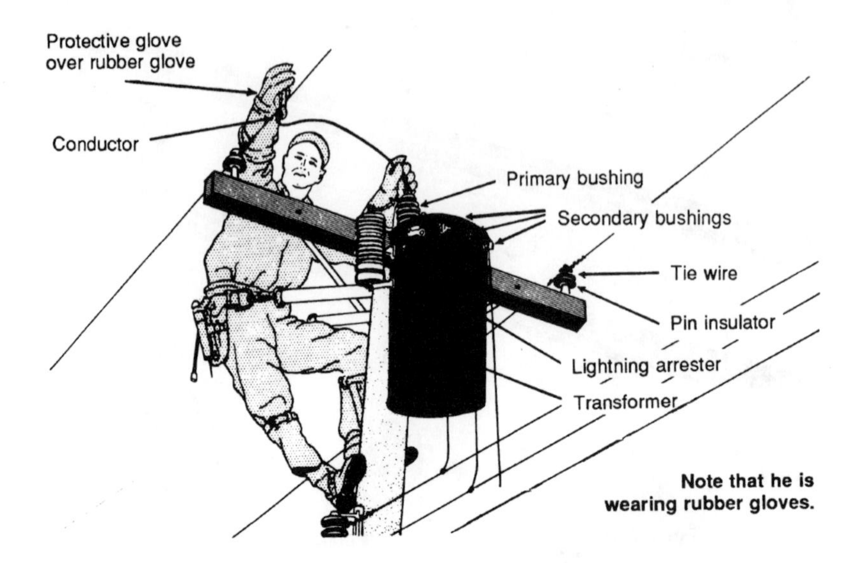

Figure 3-15. Lineworker working on line conductor.

on when energized or de-energized. They are generally situated in open areas where the danger to the public from fallen wires is negligible. Also, because the amount of insulation required would make the conductor bulky and awkward to install, it is desirable to leave high-voltage transmission line conductors bare.

Wire Sizes

In the United States, it is common practice to indicate wire sizes by gage numbers. The source of these numbers for electrical wire is the American Wire Gage (AWG) (otherwise known as the Brown & Sharpe Gage). A small wire is designated by a large number and a large wire by a small number as shown in Figure 3-16. The diameter of a No. 0000 wire is 0.4600 inch or 460 mils; the diameter of a No. 36 wire is 0.0050 inch or 5 mils. There are 38 other sizes between these two extremes. For example, a No. 8 wire is 0. 1285 inch (128.5 mils) in diameter and a No. 1 wire is 0.2576 (257.6 mils) in diameter.

It has proved convenient to discuss the cross-section area of a wire in circular mils. A circular mil (cm) is the area of a circle having a diameter of 0.001 inch or 1 mil. Because it is a circular area unit of measure, it is necessary only to square the number of mils given in the diameter of a wire to find the number of circular mils in a circle of that diameter.

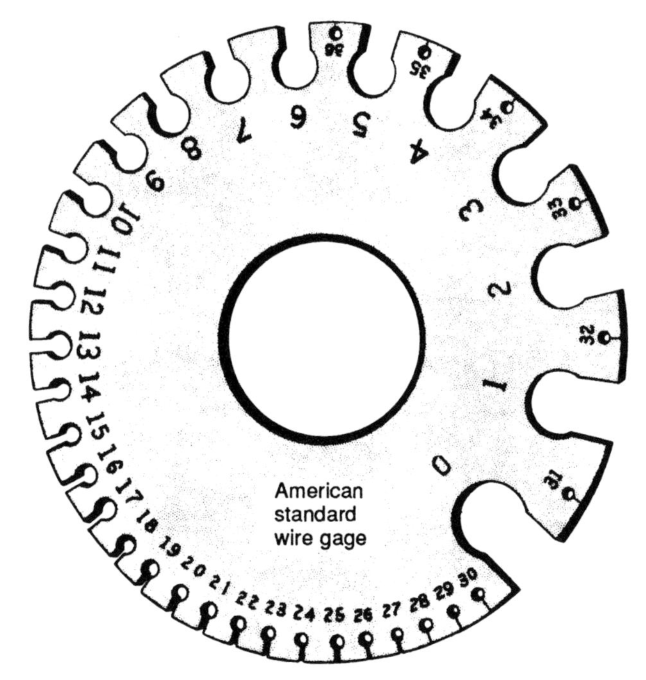

Figure 3-1. 6 American Standard Wire Gage.

Thus, a conductor with a 1-mil diameter would have a 1-circular-mil (cm) cross-section area; a 3-mil diameter wire would have a 9-cm area; and a 40-mil-diameter wire, a 1600-cm area.

For conductors larger than 0000 (4/0) in size, the wire sizes are expressed in circular mils; for example, 350,000 cm, 500,000 cm, and so on. (Sometimes these are expressed as 350 mcm, 500 mcm, etc.)

Characteristics of copper and aluminum wires are listed in Table 3-1.

Tie Wires

Conductors must be held firmly in place on the insulators to keep them from falling or slipping. On pin insulators, they are usually tied to the top or side groove of the insulator by means of a piece of wire, called a tie-wire. On suspension type insulators, conductors are usually held in place by a clamp or "shoe," as shown in Figure 3-17(a).

Where conductors must be maintained while energized, and cannot be touched by hands, they are handled on the ends of sticks called "hot sticks." When it is necessary to use these hot sticks, the ends of the tie wires are looped so that they can be easily wrapped or unwrapped

Table 3-1. Characteristics of Copper and Aluminum Wire

AWG Size	*Diameter (Mils)*	*Cross section (circular mils)*	*Resistance (ohms/1000 ft) 20°C*		*Weight (lb 1000 ft)*	
			Copper	*Aluminum*	*Copper*	*Aluminum*
0000	460	211,600	.049	0.080	640.5	195.0
00	365	133,000	.078	0.128	402.8	122.0
0	325	106,000	.098	0.161	319.5	97.0
1	289	83,700	.124	0.203	253.3	76.9
2	258	66,400	.156	0.256	200.9	61.0
3	229	52,600	.197	0.323	159.3	48.4
4	204	41,700	.249	0.408	126.4	38.4
5	182	33,100	.313	0.514	100.2	30.4
6	162	26,300	.395	0.648	79.46	24.1
7	144	20,800	.498	0.817	63.02	19.1
8	128	16,500	.628	1.03	49.98	15.2
9	114	13,100	.792	1.30	39.63	12.0
10	102	10,400	.999	1.64	31.43	9.55
11	91	8,230	1.26	2.07	24.92	7.57
12	81	6,530	1.59	2.61	19.77	6.00
13	72	5,180	2.00	3.59	15.68	4.76
14	64	4,110	2.53	4.14	12.43	3.78
15	57	3,260	3.18	5.22	9.858	2.99
16	51	2,580	4.02	6.59	6.818	2.37
17	45	2,050	5.06	8.31	6.200	1.88
18	40	1,620	6.39	10.5	4.917	1.49
19	36	1,290	8.05	13.2	3.899	1.18
20	32	1,020	10.2	16.7	3.092	0.939
35	5.62	31.5	329.0	540.0	0.0954	0.029
38	4	15.7	660.0	1080.1	0.0476	0.0145

from the insulator (see Figure 3-17(b).

Another type clamp as shown in Figure 3-17(c) is the clamp top insulator.

Connectors

Conductors are sometimes spliced by overlapping the ends and twisting the ends together, taking three or four turns. But to insure a good electrical connection as well as uniformity in workmanship, it is

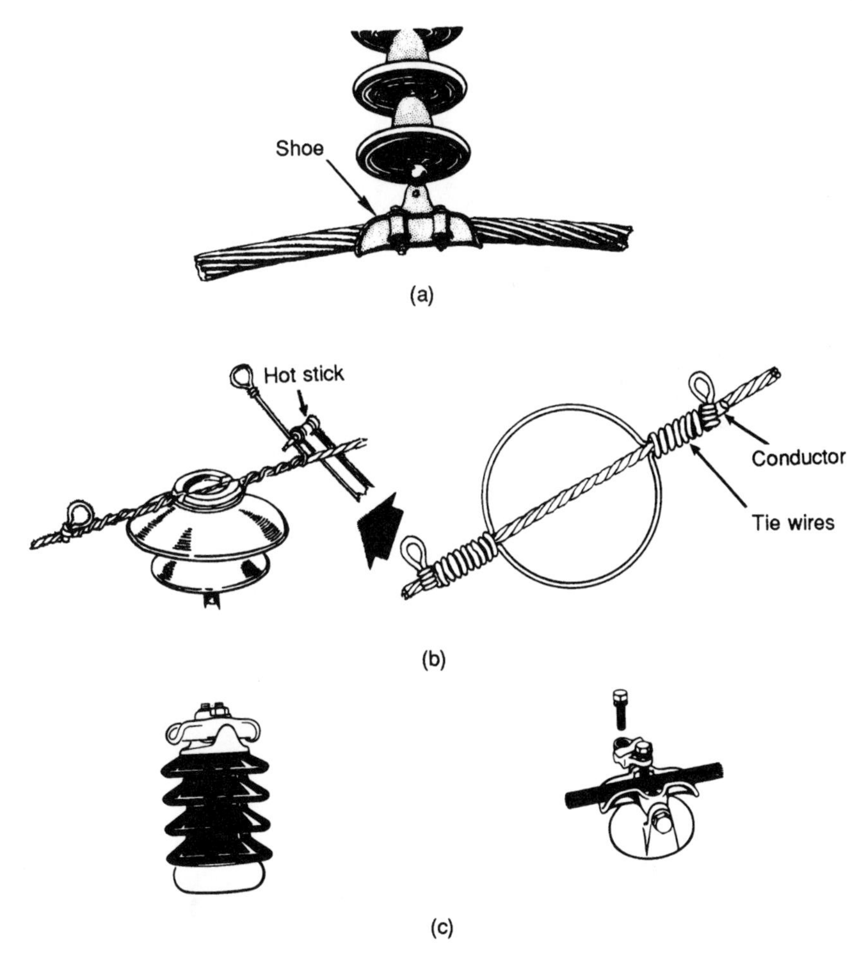

Figure 3-17. (a) Suspension insulators are attached to conductor by a shoe. (b) Tic wire for top groove of pin insulator (straightaway). (c) Clamp-top insulator.

wise to connect conductors with mechanical connectors. (Different such connectors are shown in Figure 3-18.) They are often substantial money-savers.

One type inserts the two ends into a double sleeve. When the two conductors are parallel and adjacent to each other, the sleeve is then twisted. With the compression sleeve the conductors are inserted from both ends until they butt and the sleeve is then crimped in several places. The "automatic" splice has the conductor ends inserted in each end where they are gripped by wedges held together by a spring. The split-bolt or "bug nut" connector is a copper or tin-plated-copper bolt with a channel cut into the shank; both conductors fit into the channel and are compressed together by a nut.

Another type of connector used where it is required to connect or disconnect primary conductors while energized is the "hot-line" clamp shown in Figure 3-19(a). In order to protect the conductors themselves from mechanical injury, especially from possible sparking, the clamp is often applied to a "saddle" installed on the conductor, as shown in Figure 3-19(b).

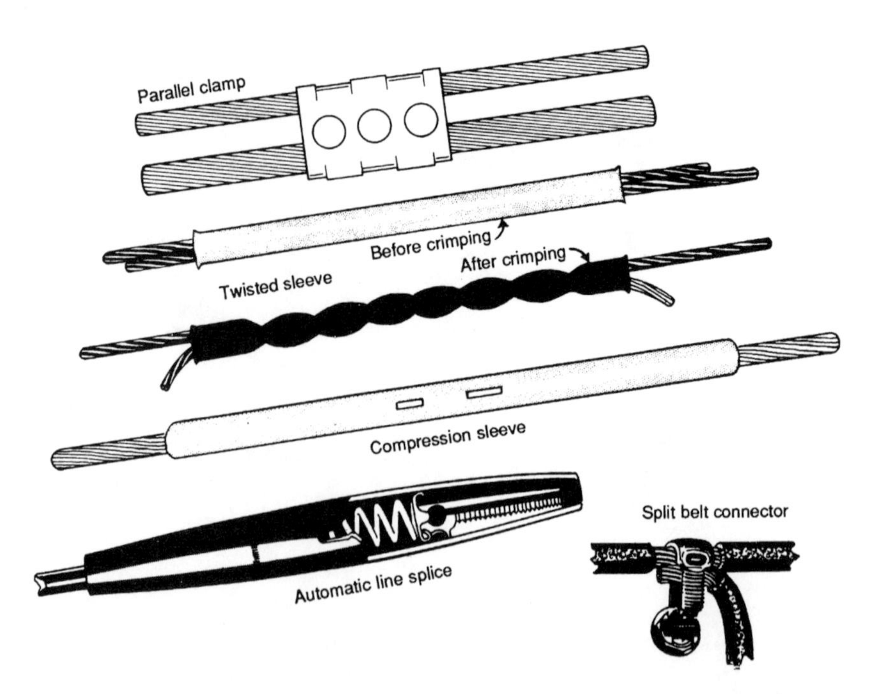

Figure 3-18. Mechanical connectors.

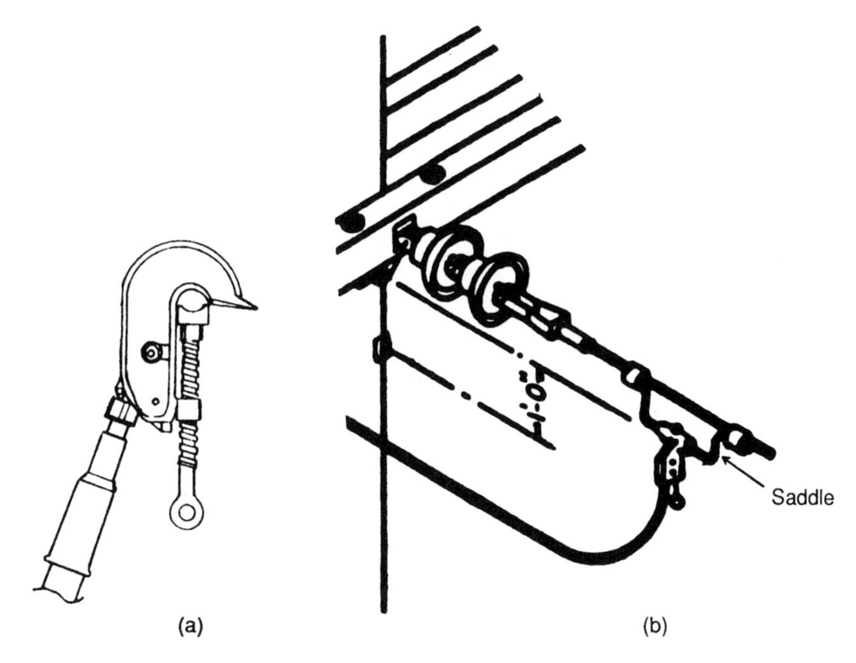

Figure 3-19. (a) Live-line or "hot-line" clamp. (*Courtesy A. B. Chance Co.*) (b) Hotline saddle. (*Courtesy Long Island Lighting Co.*)

REVIEW QUESTIONS

1. What is the function of an insulator?
2. Of what materials are insulators for overhead construction usually made? What are their electrical and mechanical properties?
3. What different types of insulators are there?
4. What are the main advantages of pin insulators? Of suspension insulators?
5. Of what materials are overhead line conductors generally made?
6. What characteristics determine a conductor's value?
7. What are the advantages of aluminum as a conductor? Steel?
8. Why are conductors stranded?
9. How are wire sizes expressed?
10. Name several types of mechanical connectors in use.

Chapter 4

Line Equipment

Besides conductors and insulators, many other pieces of equipment are necessary to get electric power from the generator to a consumer. Figure 4-1 identifies each of these pieces of equipment. Their functions will be covered in this chapter.

DISTRIBUTION TRANSFORMERS

The distribution transformer shown in Figure 4-2 is certainly the most important of these pieces of equipment. Without the distribution transformer, it would be impossible to distribute power over such long distances. Earlier in this book, we explained that the purpose of a transformer is to step up or step down voltage. In the case of the distribution transformer, the voltage is stepped down from that of the primary mains of a distribution circuit to that of the secondary mains. In most cases, this is from 2400, 4160, or 13,800 volts to 120 or 240 volts.

Most distribution transformers consist of (1) a closed-loop magnetic core on which are wound two or more separate copper coils, (2) a tank in which the corecoil assembly is immersed in cooling and insulating oil, (3) bushings for bringing the incoming and outgoing leads through the tank or cover.

Bushings

On every distribution transformer, attachments are to be found which are normally referred to as primary bushings and secondary bushings (see Figure 4-3).

A bushing is an insulating lining for the hole in the transformer tank through which the conductor must pass. Primary bushings are always much larger because the voltage is higher at that point. Sometimes the primary and secondary bushings are called high-voltage and low-voltage bushings.

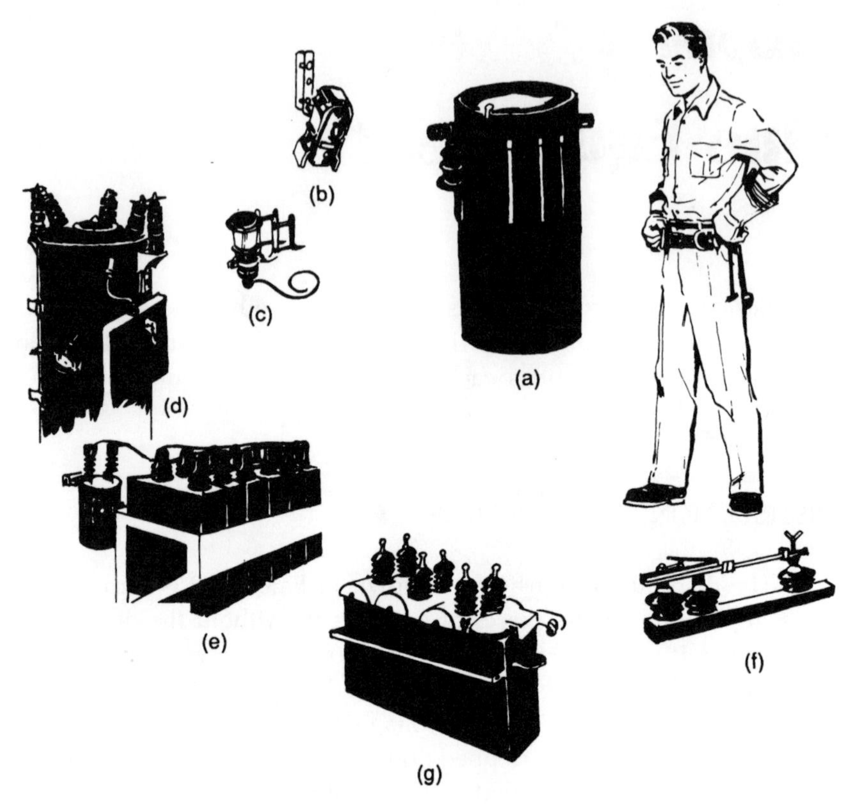

Figure 4-1. Line equipment: (a) distribution transformers, (b) fuse cutouts, (c) lightning arresters, (d) line voltage regulators, (e) capacitors, (f) switches (air and oil), (g) reclosers.

Bushings may protrude either from the sidewall of the transformer tank or from its covers. There are three types of bushings; the solid porcelain bushing, the oil-filled bushing, and the capacitor type bushing.

Solid porcelain bushings are used for voltages up to 15 kV. A solid conductor runs through the center of the porcelain form. The conductor is insulated copper cable or solid conductor terminating in a cap. The bushing cap has mountings that permit the line cables to be connected to the transformer winding. The mounting must be so designed that the cable of the transformer lead can be detached to allow the bushing to be removed for replacement.

For higher voltage transformers (discussed in the substations section), bushings are oil-filled to improve their insulating characteristics

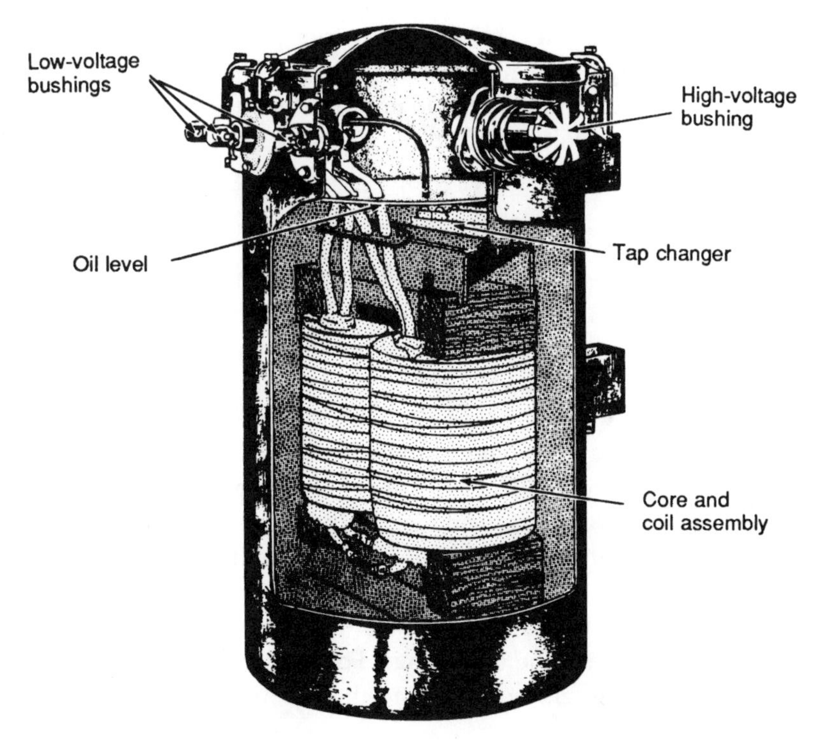

Figure 4-2. Basic components of a distribution transformer.

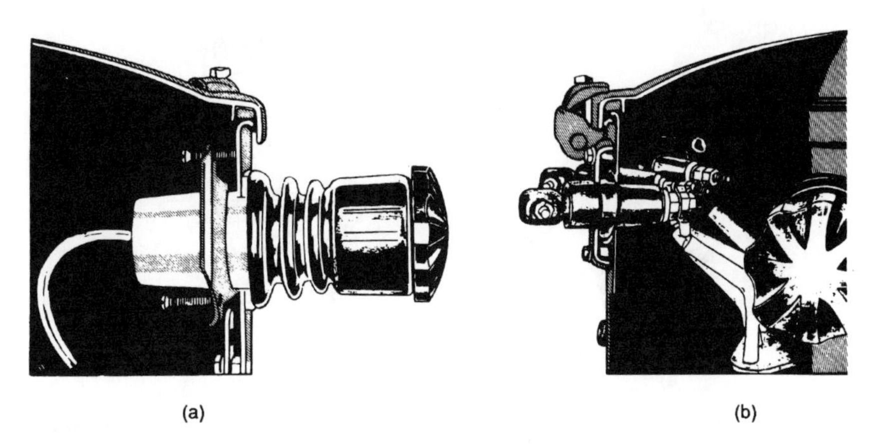

Figure 4-.3 Sidewall mounted bushings (solid porcelain). (a) Primary bushing and (b) secondary bushing.

within their specified dimensions. The interior portions of the capacitor bushing are wound with high-grade paper. The oil is replaced with paper and thin layers of metal foil to improve the distribution of stresses because of the high voltage. All three types of bushing have an outer shell of porcelain to contain the insulation inside and to shed rain.

The Tap Changer

It is often necessary to vary the voltage in a transformer winding (primary) to allow for a varying voltage drop in the feeder (transmission) lines. In other words, in spite of a varying input, the output must be nearly constant. Several ways can be used to obtain the desired result.

One method used to adjust the winding ratio of the transformer uses the no-load tap changer shown in Figure 4-4. A transformer equipped with a no-load tap changer must always be disconnected from the circuit before the ratio adjustment can be made. The selector switch is operated under oil usually placed within the transformer itself; but it is not designed to be used as a circuit breaker. To change taps on small distribution transformers, the cover must be removed and an operating handle is used to make the tap change. For the larger type, one handle may be brought through the cover and the tap may be changed with a wheel or even a motor.

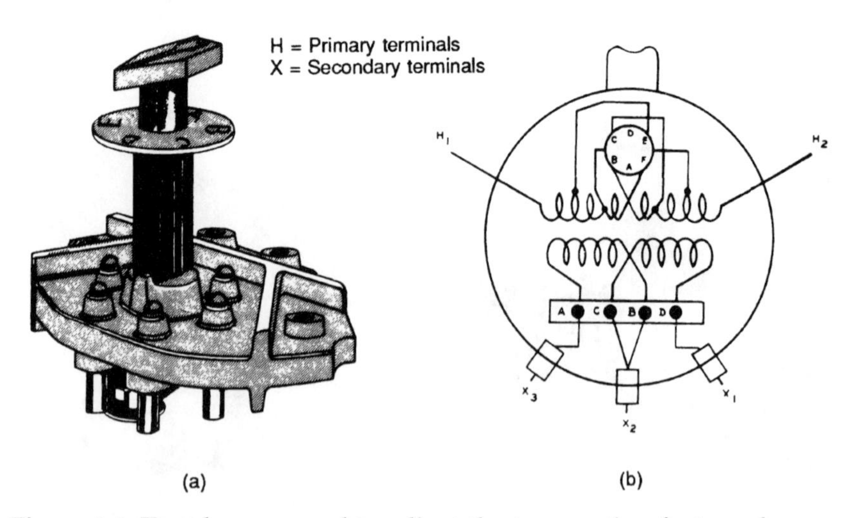

Figure 4-4. Tap changer used to adjust the turns ratio of a transformer. (a) No-load tap change and (b) typical internal wiring of transformer with tap changer.

If it is necessary to change the taps when the transformer cannot be disconnected from the circuit, tap changers under-load are used. They involve the use of an autotransformer and an elaborate switching arrangement. The information regarding the switching sequence must be furnished with each transformer. Tap changers can function automatically if designed with additional control circuits: automatic tap changes are used for high-power transformers, and for voltage regulators.

Mounting Distribution Transformers

Distribution transformers are almost always located outdoors where they are hung from crossarms, mounted on poles directly (see Figure 4-5) or placed on platforms. In general, transformers up to 75 kVA size are mounted directly to the pole or on a crossarm and larger size transformers (or groups of several transformers) are placed on platforms or mounted on poles in banks or clusters.

How a transformer is mounted is a matter of considerable importance. Remember that the distribution transformer must stay put and continue functioning even in the midst of violent winds, pouring rain, freezing cold, sleet, and snow. Besides weather, there is danger of the pole itself being hit by a carelessly driven vehicle.

Modern pole-mounted transformers [see Figure 4-6(a)] have two lugs welded directly on the case. These lugs engage two bolts on the pole as shown in Figure 46(b) from which the whole apparatus hangs securely. This method, which is known as *direct mounting* eliminates the need for crossarms and hanger irons (as was done in the past), thus saving a considerable amount of material and labor.

In the past, transformers were hung from crossarms by means of hanger irons, which were two flat pieces of steel with their top ends bent into hooks with squared sides (see Figure 4-7). The transformer was bolted to these pieces of steel, the assembly was raised slightly above the crossarm and then lowered so that the hooks on the hanger irons would engage the crossarm. Many such installations still exist.

A transformer should not be mounted on a junction pole (a pole supporting lines from three or more directions) as this makes working on such a pole more hazardous for the worker.

Where transformers cannot be mounted on poles because of size or number, they may be installed on an elevated platform [Figure 4-8(a)] or a ground-level pad [Figure 4-8(b)]. Platforms are built in any shape or size required to suit the particular need. They are usually constructed of

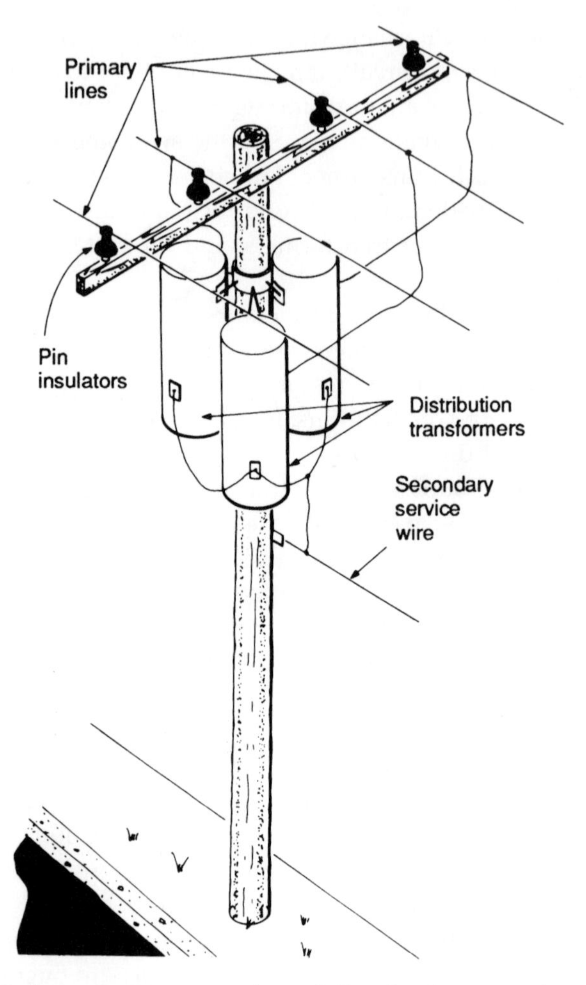

Figure 4-5. Modern cluster mounting of distribution transformers on a pole.

wood, though steel is often used for some of the members. Ground-level pads are usually made of concrete or reinforced concrete and are very useful when appearance is a major consideration.

A Conventional Distribution Transformer

Manufacturers produce two types of distribution transformers: the conventional and the completely self-protected. (CSP is the Westinghouse trade designation for this latter type. Other manufacturers

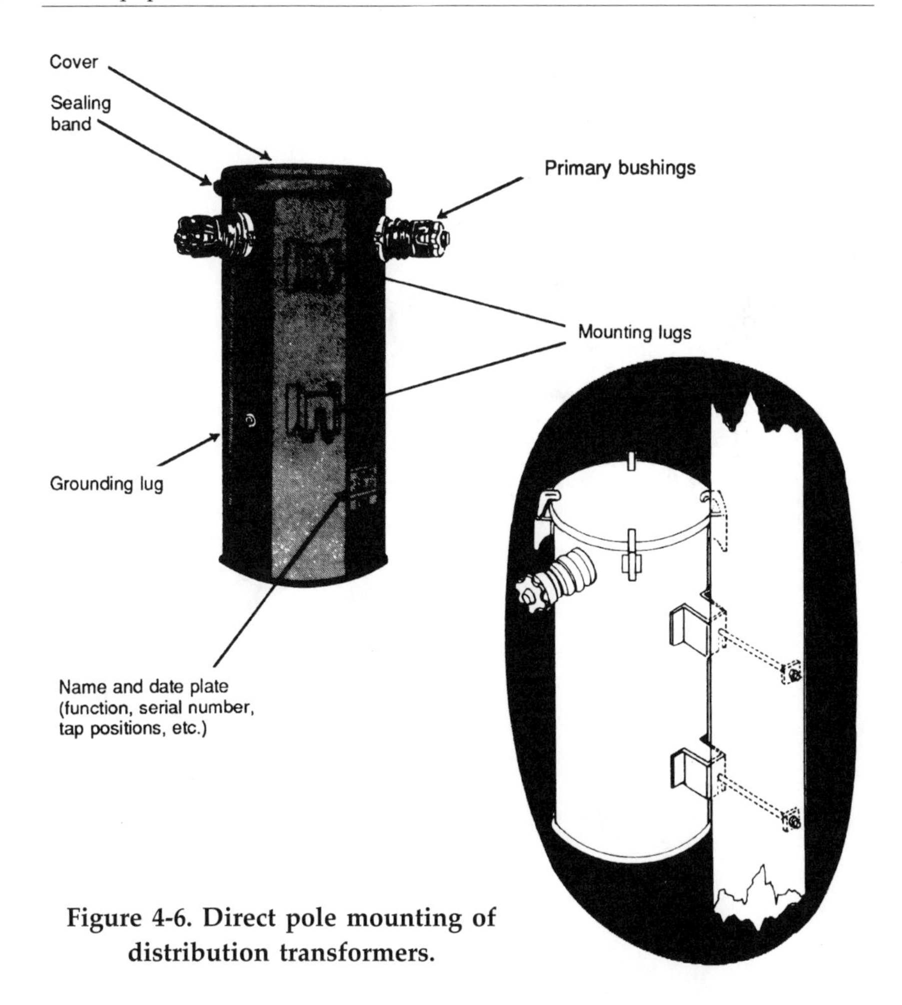

Figure 4-6. Direct pole mounting of distribution transformers.

simply call them self-protected transformers.) A conventional distribution transformer consists only of a case containing the transformer unit with protective devices, usually a primary fuse cutout and a lightning arrester, mounted separately on the pole or crossarm (see Figure 4-9).

CSP Distribution Transformers

In the CSP transformer, a *weak link* or primary protective *fuse* link shown in Figure 4-10 is mounted inside the tank with the transformer unit as also are two circuit breakers for protection on the secondary side of the transformer. A simple thermal device causes the breakers to open when a predetermined safe value of temperature is exceeded. The light-

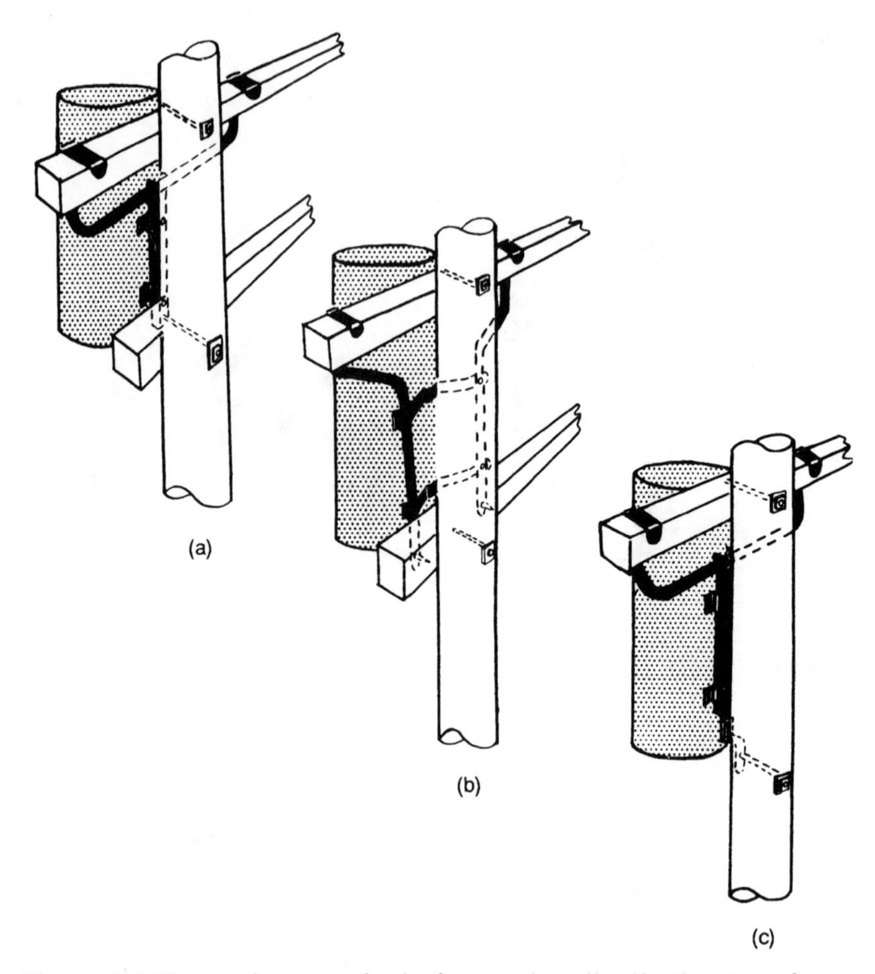

Figure 4-7. Hanger-iron method of mounting distribution transformers on poles. (a) T-shaped hanger iron using two crossarms. (b) C-shaped hanger-iron designed to hold larger transformers. (c) For smaller transformers, a T-shaped hanger-iron one crossarm and a kicker are sufficient.

ning arrester is mounted on the outside of the tank. It is apparent that the CSP transformer makes for simpler, more economic mounting and neater appearance. What's more, it is of particular advantage for higher voltage (13.8 kV) primary distribution systems where connections and disconnections are made with hot sticks.

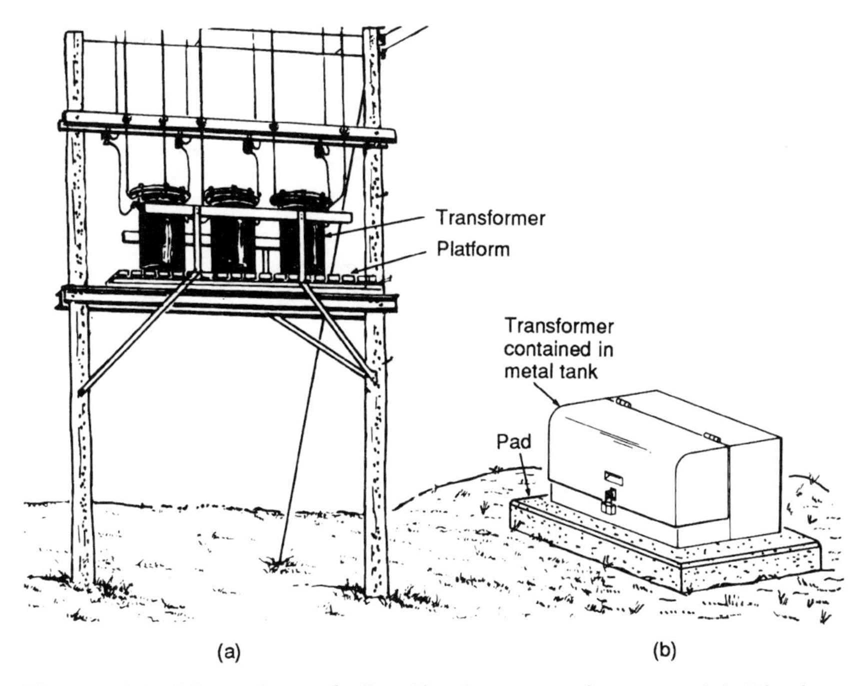

Figure 4-8. Mounting of distribution transformers. (a) Platform mounted and (b) pad mounted.

FUSE CUTOUTS

Suppose a circuit is designed to carry 100 A. Should the amperage climb over that limit, it could eventually melt some of the wires and cause widespread interruption of service. To prevent this, a weak spot is intentionally designed into a circuit—a place where overload will register and open the circuit almost immediately. This spot is called *a fuse*. A fuse consists of a short piece of metal having low melting characteristics which will melt at a rated temperature.

It is amperage flowing through a conductor that sometimes makes the conductor hot to touch. This is what the fuse counts on. Should there be an overload, the fuse melts, thus disconnecting the circuit.

Fuse cutouts go one step further. They may be placed as shown in Figure 4-11 so as to cut out the section of the circuit that is endangered, allowing the rest of the circuit to remain energized.

A primary fuse cutout is connected between the primary lines and

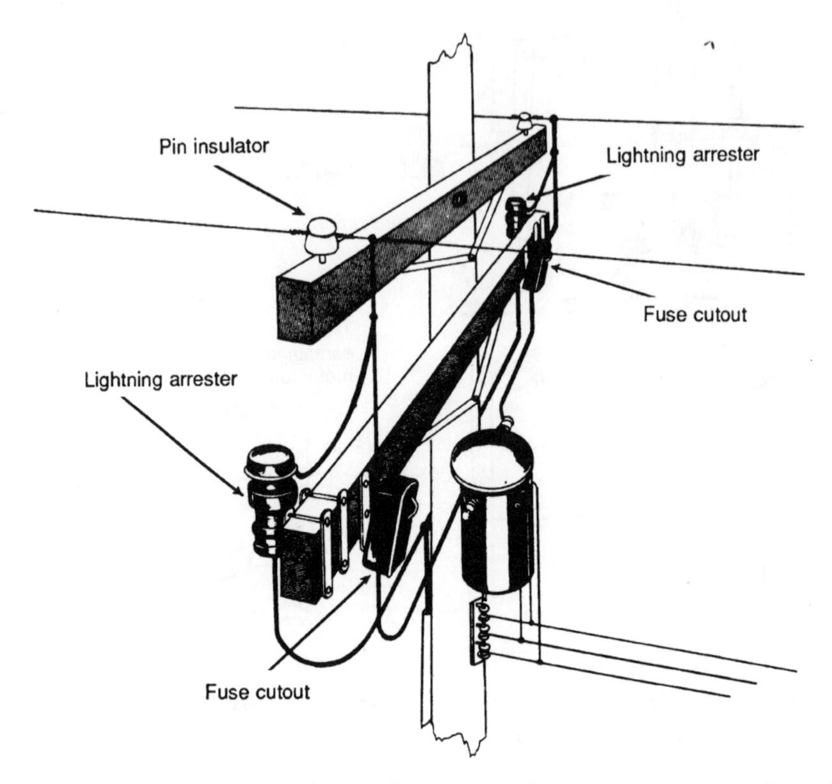

Figure 4-9. A conventional transformer requires separate mounting of lightning arresters and fuse cutouts.

the transformer shown in Figure 4-12 to protect the transformer from overloads and to disconnect the transformer from the primary lines in case of trouble. A primary line fuse cutout can disconnect any portions of a primary circuit supplying several distribution transformers, in case of overload or fault, leaving the rest of the circuit energized.

Although there are several types of fuse cutouts, the principle upon which each is constructed is the same. A fuse ribbon makes a connection between two contacts, either the line and the transformer or the main line and that portion to be protected.

The Door-type Cutout

In the door or box-type cutout shown in Figure 4-13, the fuse is mounted inside the door in such a manner that when the door closes, the fuse engages two contacts, one on the top of the box and the other

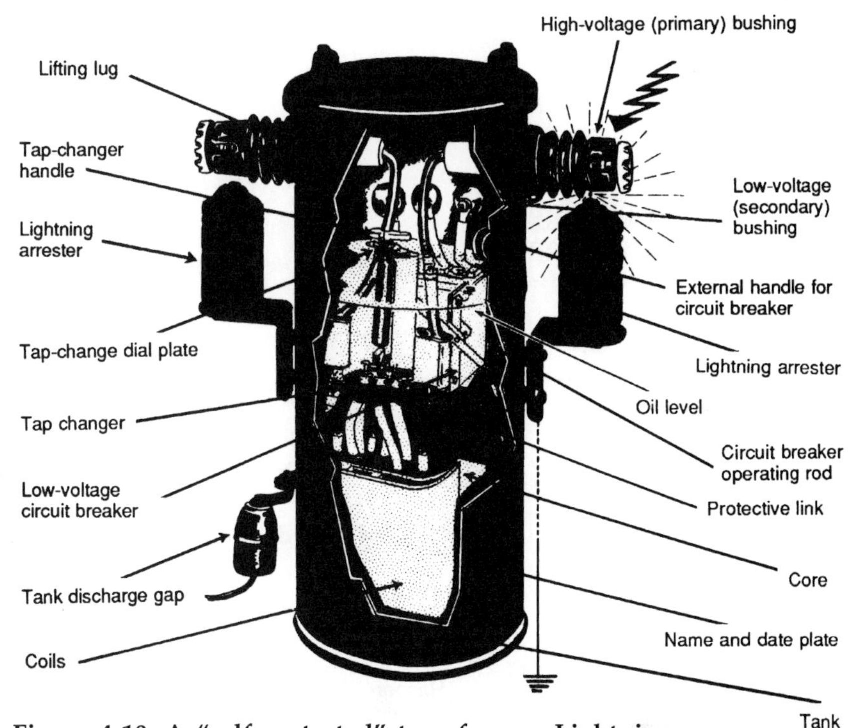

Figure 4-10. A "self-protected" transformer. Lightning arresters and circuit breaker are integral parts of its design.

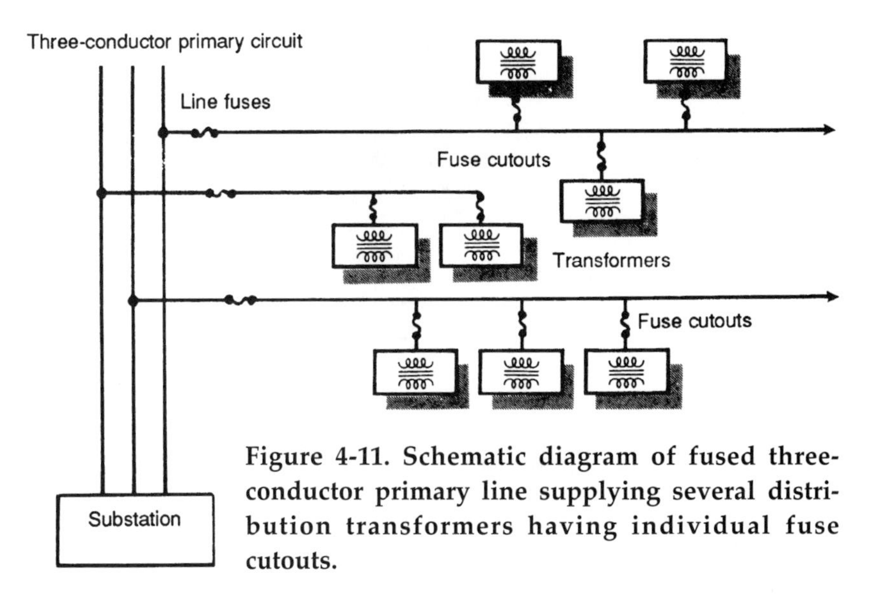

Figure 4-11. Schematic diagram of fused three-conductor primary line supplying several distribution transformers having individual fuse cutouts.

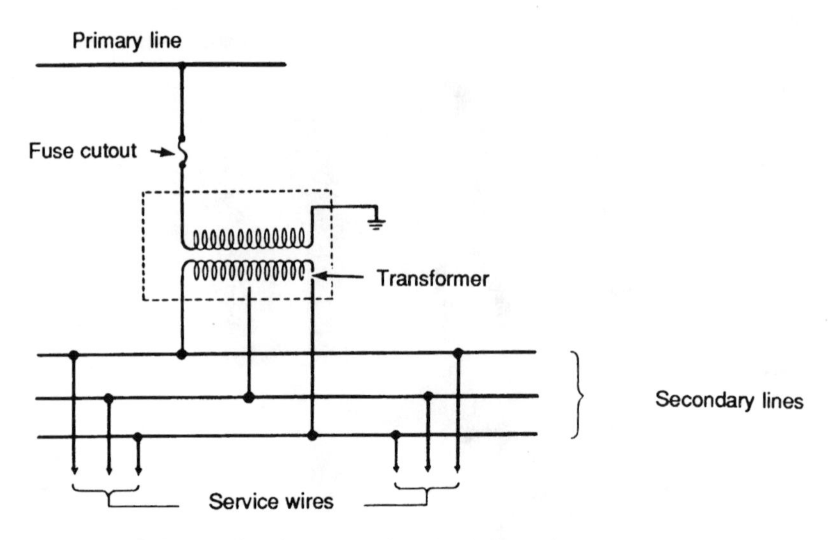

Figure 4-12. Schematic diagram of a single-conductor primary line employing a fused transformer.

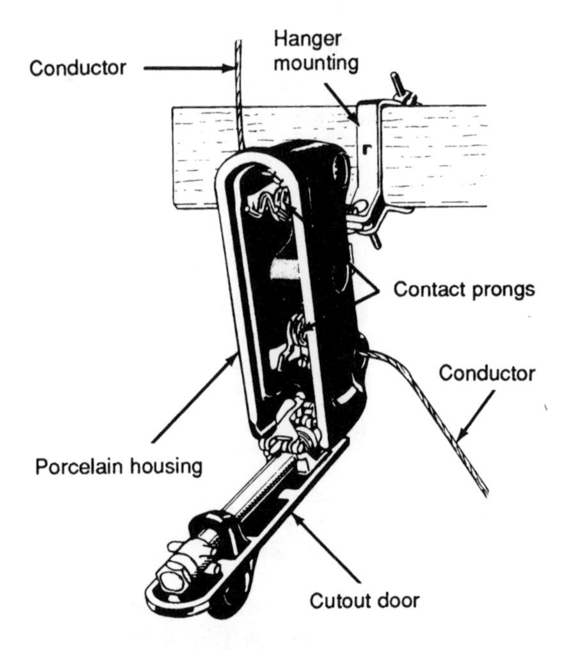

Figure 4-13. Box-type cutout. Open door indicates fuse has blown.

on the bottom. To open this cutout, the door is pulled open and allowed to hang downward from the box. The fusible element is enclosed in a fiber tube; when the fuse blows or melts because of excessive current passing through it, the resultant arc attacks the fiber tube, producing a gas which blows out the arc. For this reason this type cutout is sometimes also called the expulsion type. In later models, the fuse mounting is arranged so that the melting of the fusible element causes the door to drop open, signaling to the worker on the ground that the fuse has blown. The use of this type cutout is generally limited to circuits operating at voltages under 5000 volts.

The Open-type Cutout

The open-type cutout (Figure 4-14) is essentially the same as the door type, except that the fiber tube enclosing the fusible element is exposed in the open, rather than enclosed in a porcelain box. This arrangement enables larger currents to be interrupted without confining as much the attendant violent expulsion of gases (which can destroy the cutout). The tube drops when the fuse blows, indicating that the fuse has blown. This type cutout is used on distribution circuits operating at voltages over 5000 volts, although it can be used on lower voltage circuits.

The Repeater Fuse

When line fuses are used to protect a portion of a primary circuit as previously described, a repeater fuse may be used. The repeater fuse shown in Figure 4-15 is usually of the open type and consists of two or three fuses mechanically arranged so that when the first fuse blows and drops, the action places the second fuse automatically in the circuit. If the trouble has been cleared, service will be restored. Should the second fuse also blow, a third is also automatically connected in the circuit; when the third fuse blows, the portion of the circuit is finally de-energized. Repeater fuses hold down to a minimum interruptions in service caused by temporary faults. These faults may arise from wires swinging together when improperly sagged, or from tree branches or animals making momentary contact with the line, or from lightning surges causing temporary flashover at an insulator.

All of these fused cutouts are mounted on the crossarm or on the pole by bolts and a steel bracket.

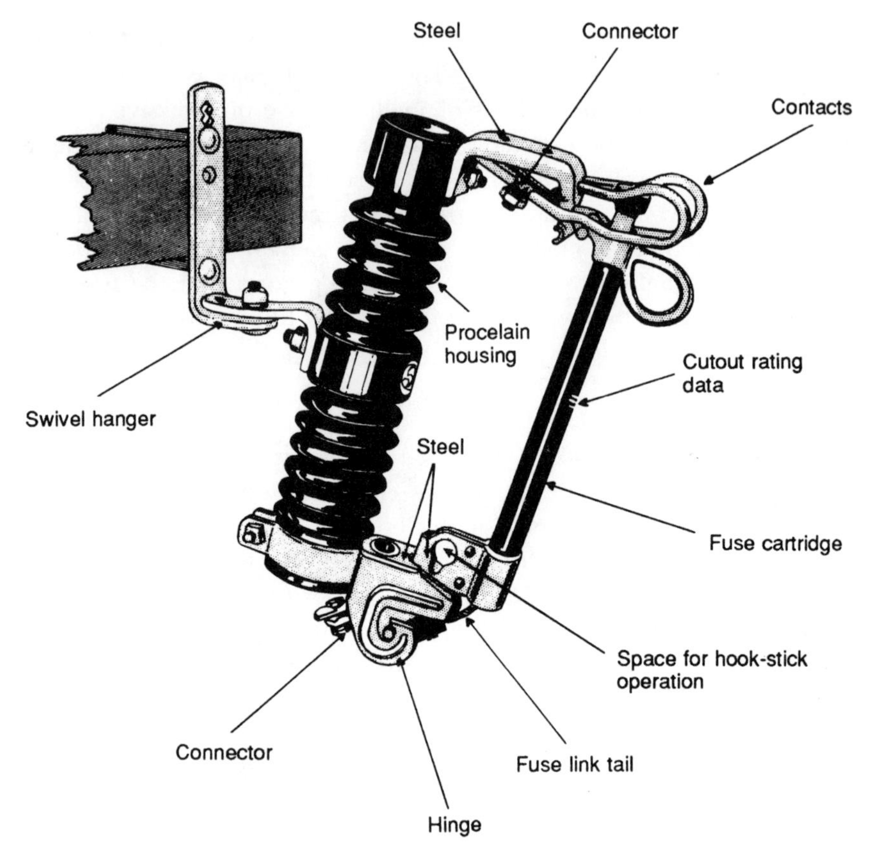

Figure 4-14. Open-type fuse cutout.

LIGHTNING OR SURGE ARRESTERS

A lightning arrester is a device that protects transformers and other electrical apparatus from voltage surges. These surges can occur either because of lightning or improper switching in the circuit. The lightning arrester provides a path over which the surge can pass to ground as shown in Figure 4-16(a), before it has a chance to attack and seriously damage [see Figure 4-16(b)] the transformer or other equipment.

The elementary lightning arrester consists of an air gap (illustrated in Figure 4-17) in series with a resistive element. The voltage surge causes a spark which jumps across the air gap, passes through the resistive element (silicon carbide, for example) which is usually a material

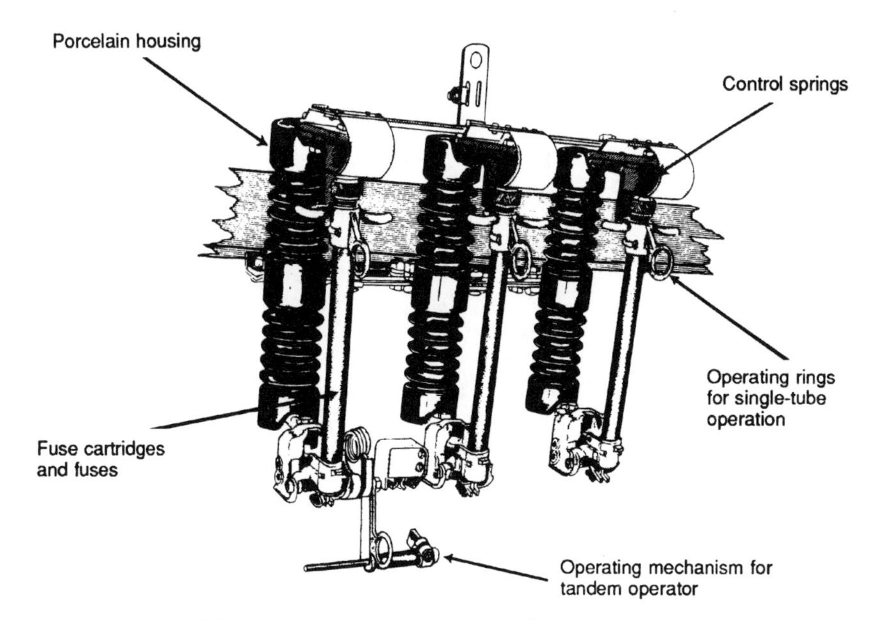

Figure 4-15. Repeater-type fuse cutout.

that allows a low-resistance path for the high-voltage surge, but presents a high resistance to the flow of line energy. This material is usually known as the "valve" element. There are many different types of lightning arresters, but they generally have this one principle in common. There is usually an air gap in series with a resistive element, and whatever the resistive (or valve) element is made of, it must act as a conductor for high-energy surges and also as an insulator toward the line energy. In other words, the lightning arrester leads off only the surge energy. Afterwards, there is no chance of the normal line energy being led into ground.

Valve Arresters

Since all arresters have a series gap and a resistive element, they differ only in mechanical construction and in the type of resistive element used. One type arrester consists of a porcelain cylinder filled with some suitable material with electrodes at either end (see Figure 4-18). The series gap assembly is usually at the top. When there is excessive voltage surge on the line, the spark jumps across the air gap and the surge energy flows through the material to the ground. As the surge

Figure 4-16. Lightning arresters protect transformers and other equipment. (a) With arrester, surge voltage is drained to ground. (b) With no arrester, surge voltage causes damage.

decreases, the resistive power of the material increases so that no line energy will flow to the ground.

Expulsion-type Arresters

In this type arrester there are two series gaps (shown in Figure 4-19)—one external and one internal—which are to be bridged by the high voltage surge. The second gap formed between two internal electrodes is actually inside a fiber tube which serves to quench the line energy

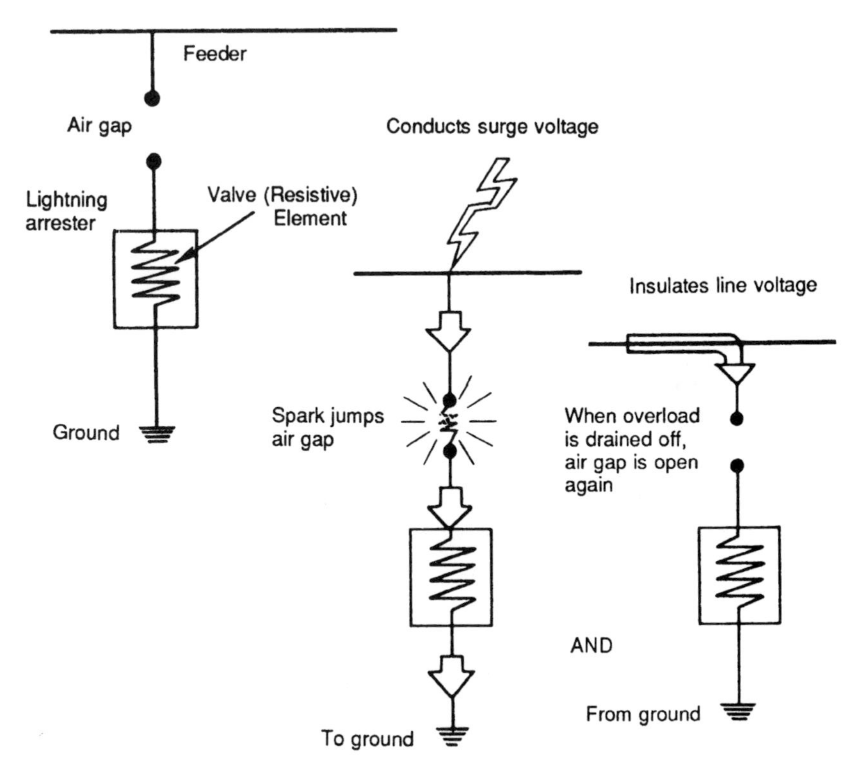

Figure 4-17. Elementary lightning arrester.

when it comes through. (Thus, the valve element in this case is the fiber tube itself.) When lightning occurs, the gaps are bridged and the high energy flows harmlessly to ground. However, when the line energy tries to go through the arcing channel, the fiber tube creates nonconducting gases which in turn blow the arc and conducting gases out the vent, thus reestablishing a wall of resistance to the line energy.

VOLTAGE REGULATORS

At this point, voltage regulators for pole-mounting will be discussed. Later, when substations are discussed, these will be used there as well.

A voltage regulator is generally used to maintain the voltage of a line. The primary feeder voltage generally drops when a large load current is drawn and less voltage is available across the primaries of the

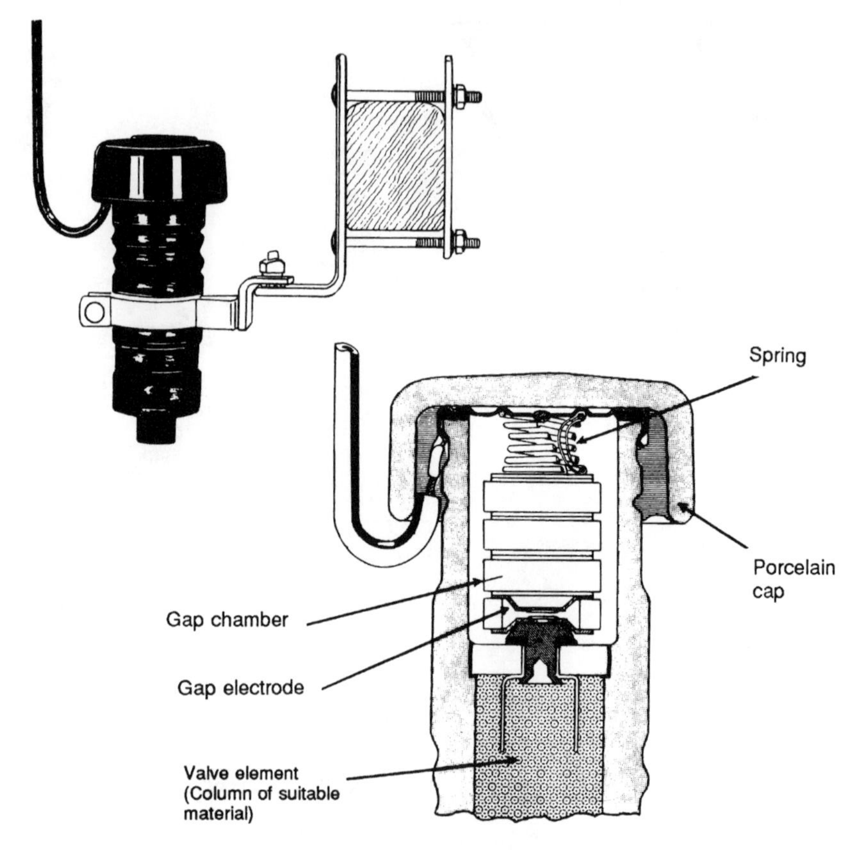

Figure 4-18. Valve lightning arrester.

distribution transforms. The regulator maintains the voltage at the proper rated value at all times.

The principle of operation of a voltage regulator is somewhat similar to that of a transformer having taps, as previously described. This form of regulator has two fixed windings, a primary (high-voltage) winding connecting in shunt or across a line, and a secondary or low-voltage winding connected in series with the line. The secondary or series winding is provided with as many taps as necessary (shown in Figure 4-20) to vary the voltage across this winding. This equipment operates as a voltage regulator by means of a control circuit which automatically changes the tap setting on the series winding, while leaving the voltage applied to the primary (high-voltage) winding alone. The variable voltage in the series winding can thus be added or subtracted

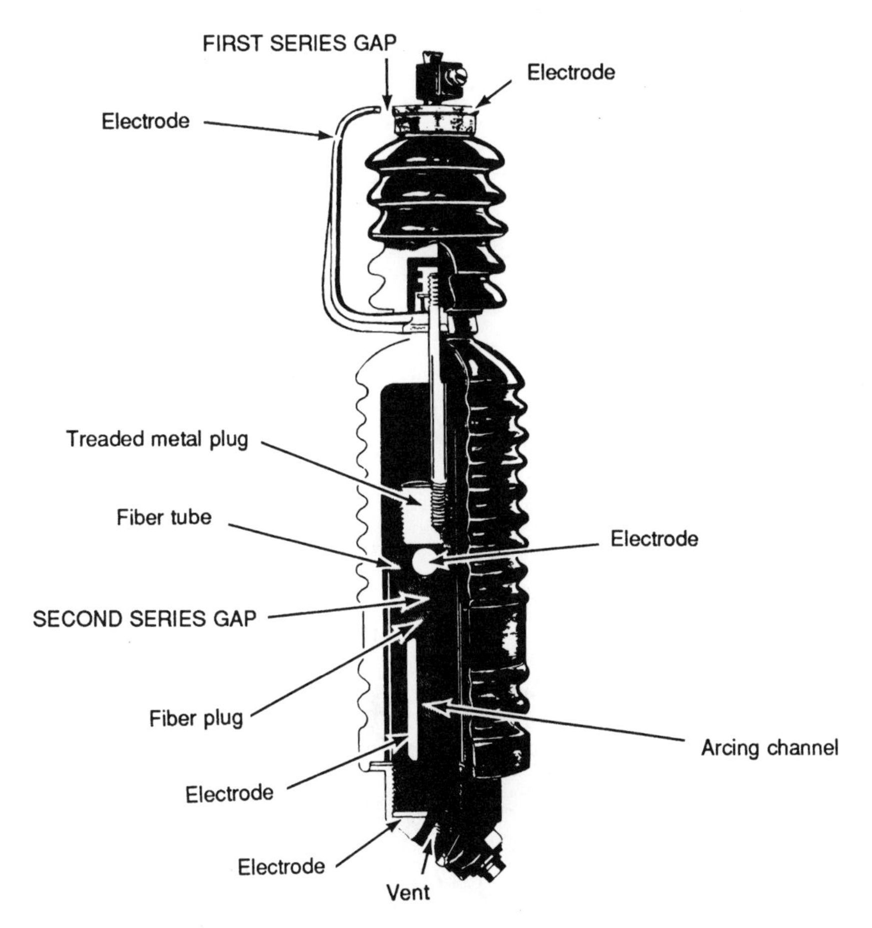

Figure 4-19. Expulsion lightning arrester.

from the incoming (or primary) voltage, resulting in an outgoing voltage which can be kept approximately constant even when the incoming primary voltage may vary.

Another type, known as the induction-type voltage regulator, accomplishes the same effect by having the primary coil rotate, changing its position in relation to the secondary coil, which in this case has no taps. (Figure 4-21)

Voltage regulators are either hand or motor-operated. When a motor is used, it is usually automatically controlled by means of relays.

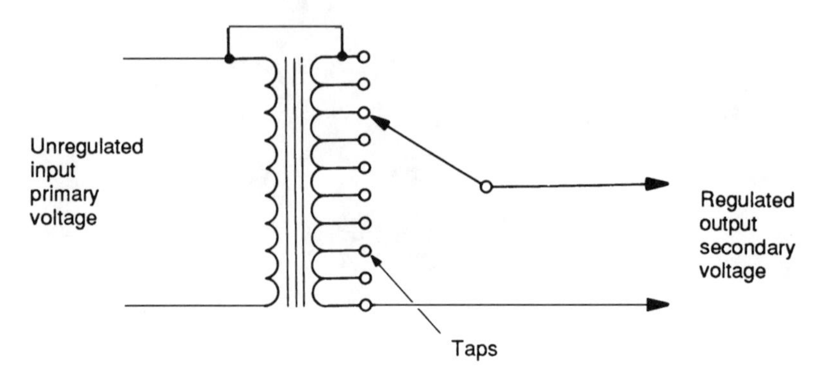

Figure 4-20. Voltage regulator. Schematic diagram of tap-changing-under load (TCUL) regulator.

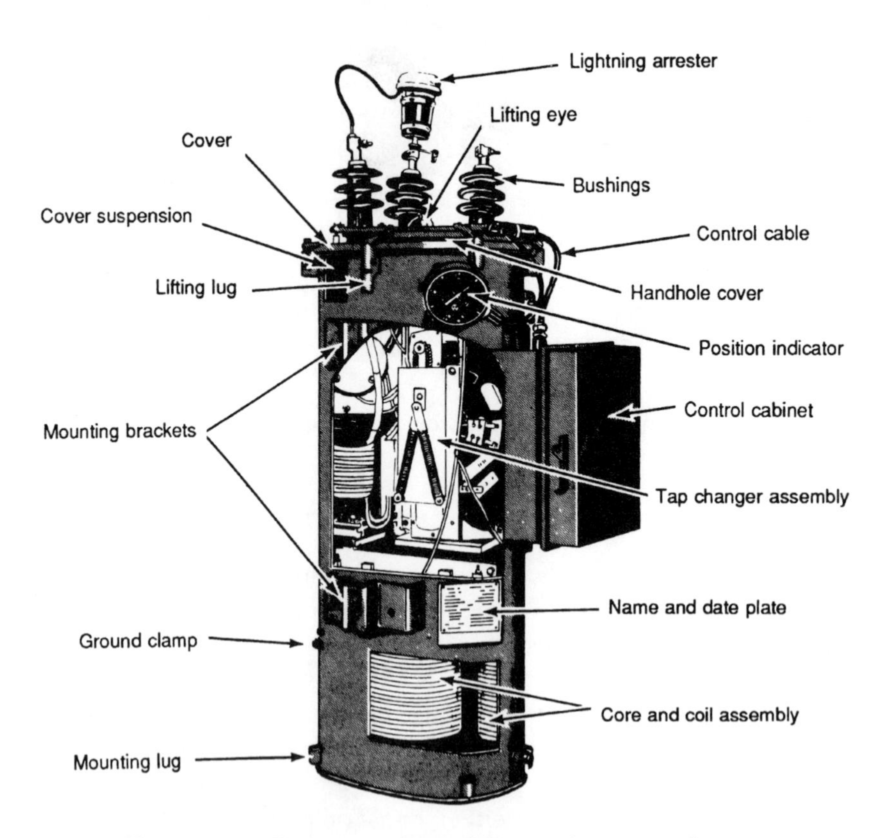

Figure 4-21. Cutaway view of line-voltage regulator.

CAPACITORS

Should the voltage on a circuit fall below a specified level for some reason, a device called a capacitor can momentarily maintain the voltage at line value. Basically, a capacitor serves the same purpose as a storage tank in a water system. By maintaining the water in a storage tank at a definite level, the pressure on the water supplied by the system connected to it is maintained evenly.

It is the job of capacitors to keep the *power factor* as close to 1 as possible. The power factor is an important essential of electricity which will be explained later. At this point, let it suffice to say that keeping the power factor close to 1 is a considerable economic advantage to the utility company and to the consumer. Inductance is the element in the circuit which is pulling the power factor below 1. Capacitance is the enemy of inductance. Therefore, capacitors counteract inductance, keep the power factor close to 1, and save money for the utility company.

The capacitor usually consists of two conductors separated by an insulating substance. Among other materials which may be used, a capacitor can be made of aluminum foil separated by oil-impregnated paper (see Figure 4-22), or synthetic insulating materials.

Capacitance is the property of a capacitor. Capacitance depends on the area of the conductors, on the distance between the conductors and on the type of insulating material used.

Introducing capacitors into a circuit causes the current to lead the voltage in phase. Introducing inductance (or an inductor) into a circuit causes the current to lag the voltage in phase. In most power applications, inductance prevails and reduces the amount of pay-load power produced by the utility company for a given size of generating equipment. The capacitor counteracts this loss of power and makes power-production more economical.

Capacitors are mounted on crossarms or platforms (see Figure 4-23) and are protected with lightning arresters and cutouts, the same as transformers. Figure 4-24 illustrates the many uses that are made of capacitors.

SWITCHES

Switches shown in Figure 4-25 are used to interrupt the continuity of a circuit. They fall into two broad classifications: air switches and oil

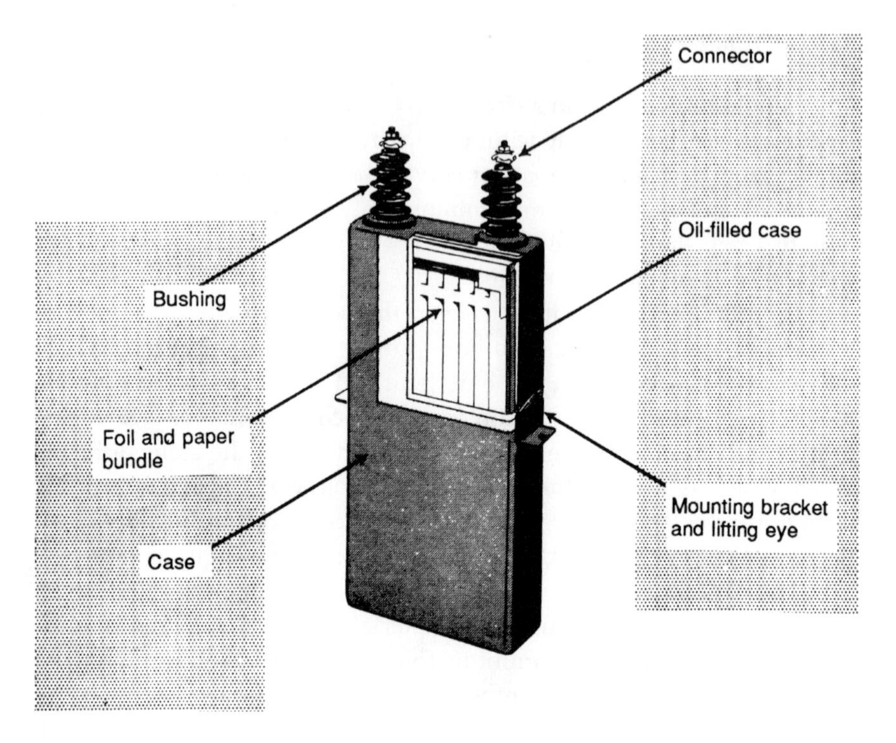

Figure 4-22. A primary capacitor.

vacuum or gas (SF_6) switches. As their names imply, air switches are those whose contacts are opened in air, while the other type switches are those whose contacts are opened in oil, vacuum, or gas. Oil switches are usually necessary only in very high-voltage, high-current circuits.

Air switches are further classified as air-break switches and disconnect switches.

Air-break Switches

The air-break switch shown in Figure 4-26 has both the blade and the contact equipped with *arcing horns.* These are pieces of metal between which the arc resulting from opening a circuit carrying current is allowed to form. As the switch opens, these horns are spread farther and farther apart and the are is lengthened until it finally breaks.

Air-break switches are of many designs. Some are operated from the ground by a hook on the end of a long insulated stick; some others through a system of linkages are opened by a crank at the foot of the

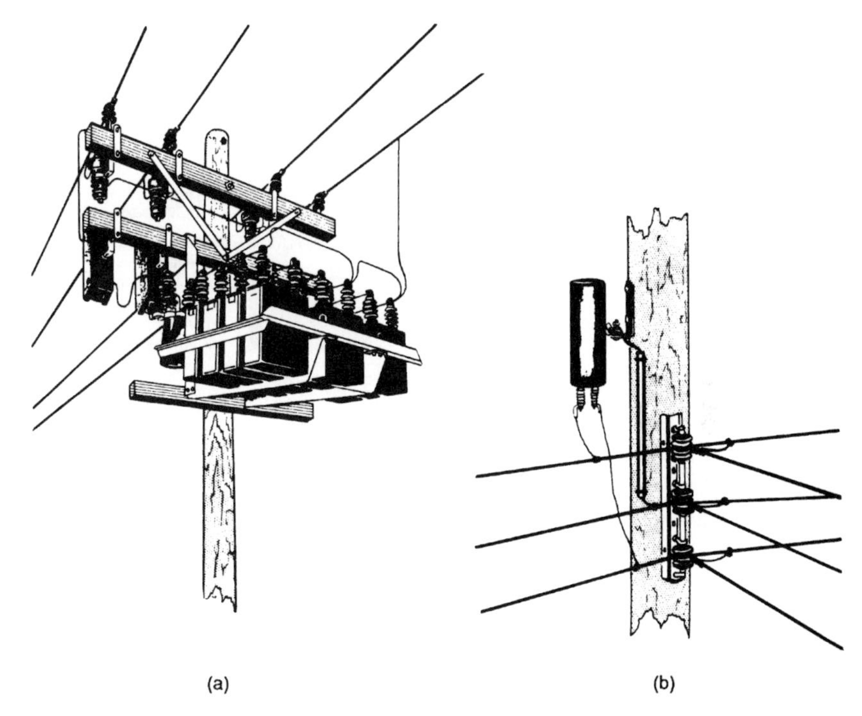

Figure 4-23. Pole-mounted capacitors. (a) Primary and (b) secondary.

pole. Where more than one conductor is opened, there may be several switches mounted on the same pole. These may be opened singly or altogether in a "gang" as this system is called. Some switches are mounted so that the blade opens downward and these may be provided with latches to keep the knife blade from jarring open.

A modern development of the air-break switch is the load-break switch shown in Figure 4-27 which breaks the arc inside a fiber tube. As in the expulsion lightning arrester, the fiber tube produces a gas which helps to confine the arc and blow it out. There is a possibility that the unconfined are associated with the horn-type switch might communicate itself to adjacent conductors or structures causing damage and possible injury. But the load-break switch eliminates this potential hazard.

The important element of the load-break switch is the interrupter unit shown in Figure 4-27. Naturally, the heart of the unit is the arc-extinguishing section which consists of a pair of arcing contacts (one stationary and one movable) and a trailer operating within a fiber bore.

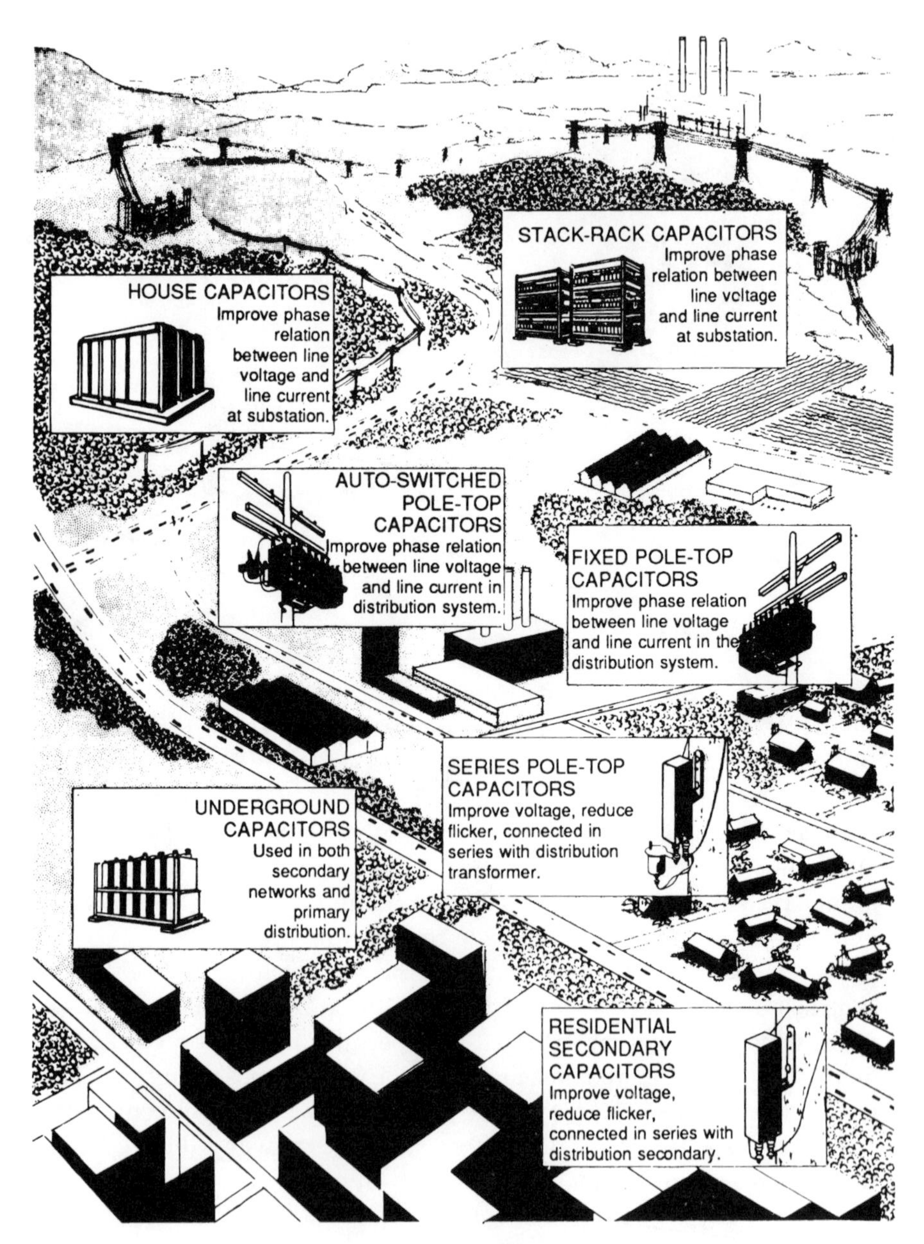

Figure 4-24. How capacitors are used.

The trailer which is made of acrylic resin, follows the contact through the bore, confining the arc between the fiber wall and itself. The arc is extinguished by deionizing gases coming from both the fiber and the acrylic resin.

The disconnect switch (Figure 4-28) is not equipped with arc-quenching devices and, therefore, should not be used to open circuits carrying current. This disconnect switch isolates one portion of the circuit from another and is not intended to be opened while current is flowing. Air-break switches may be opened under load, but disconnect switches must not be opened until the circuit is interrupted by some other means (see Figure 4-29).

Oil Switches

The oil switch has both the blade and the contact mounted in a tank filled with oil. The switch is usually operated from a handle on the

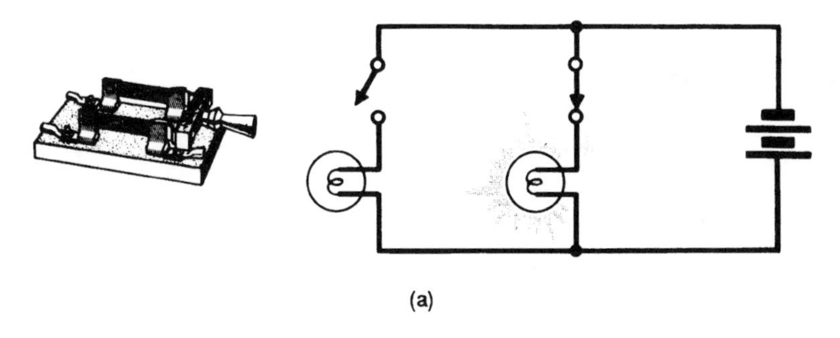

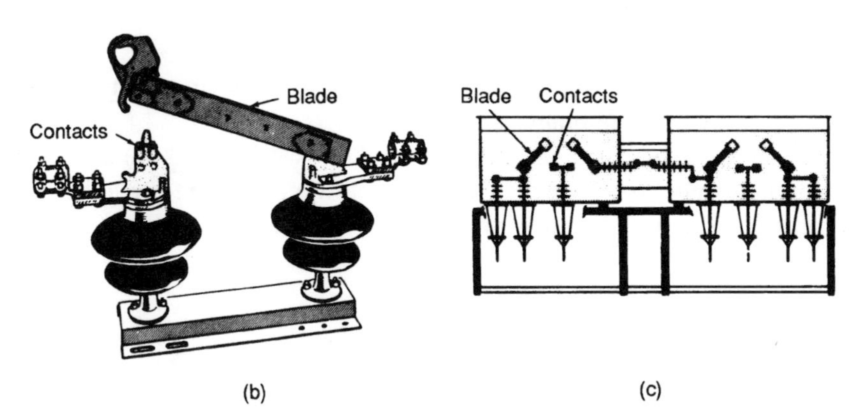

Figure 4-25. Switches interrupt the continuity of a circuit. (a) Typical switch, (b) air switch, and (c) oil switches.

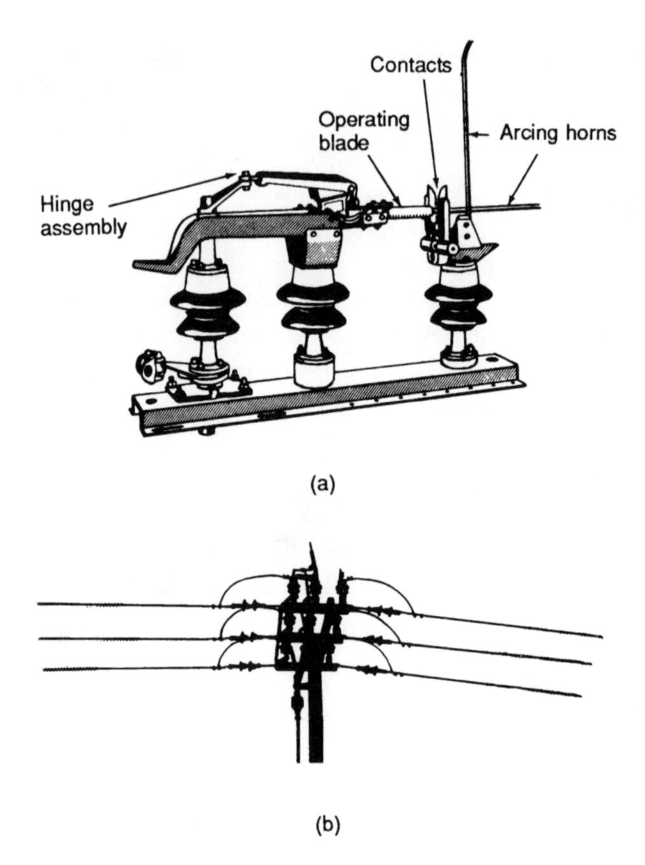

Figure 4-26. Air-break switch in closed position (with arcing horns) (a), and pole-mounted air-break switch in open position (b).

outside of the case. As the switch opens, the arc formed between the blade and contact is quenched by the oil.

Oil switches may be remote controlled (see Figure 4-30) as well as manually operated. They are used for capacitor switching, street lighting control and automatic disconnect in case of power failure.

RECLOSERS

Oil-circuit Reclosers

A recloser consists essentially of an oil switch or breaker actuated by relays which cause it to open when predetermined current-values

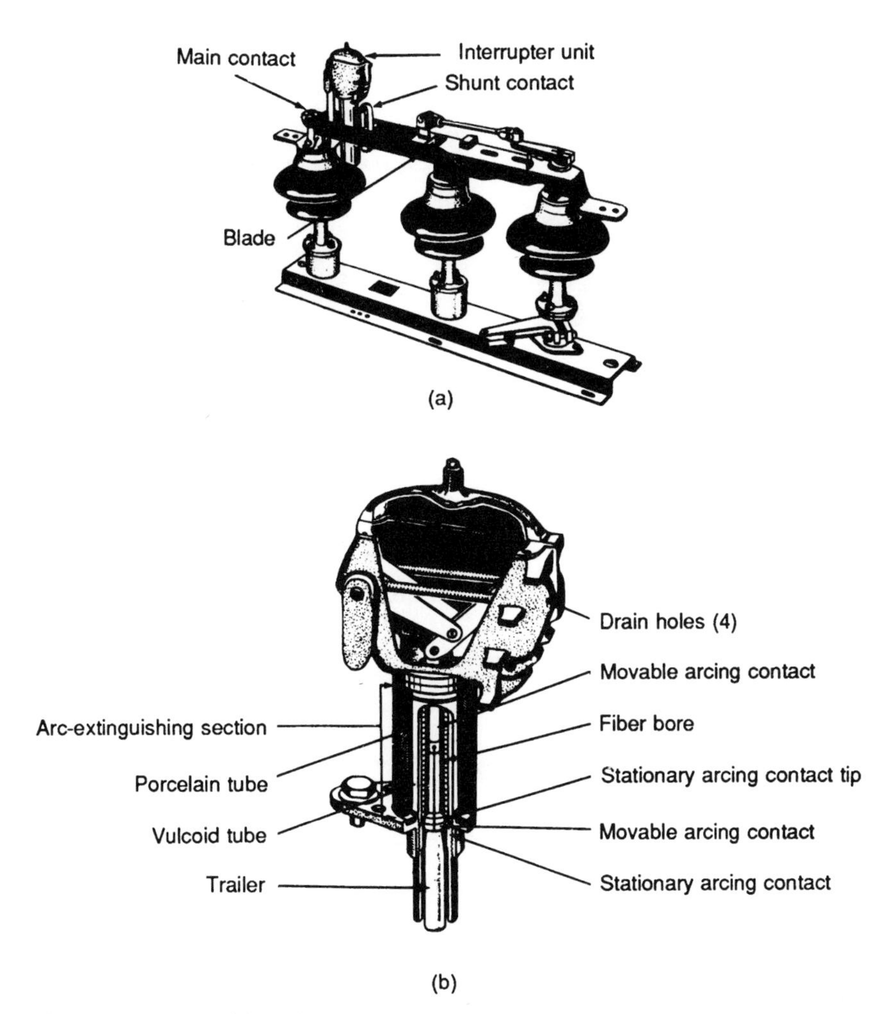

Figure 4-27. Load-break air-switch (a), and cutaway view of the interrupter unit (b).

flow through it. The recloser resembles the repeater fuse cutout described previously In many ways. Reclosers are usually connected to protect portions of primary circuits and may take the place of line fuses. The switch or breaker is arranged to reclose after a short interval of time and re-open again should the fault or overload which caused the excess current-flow persist. Also, the same as the repeater fuse cutout, the recloser can be set for three or four operations before it locks itself open for manual operation. It differs from the repeater fuse cutout in that its

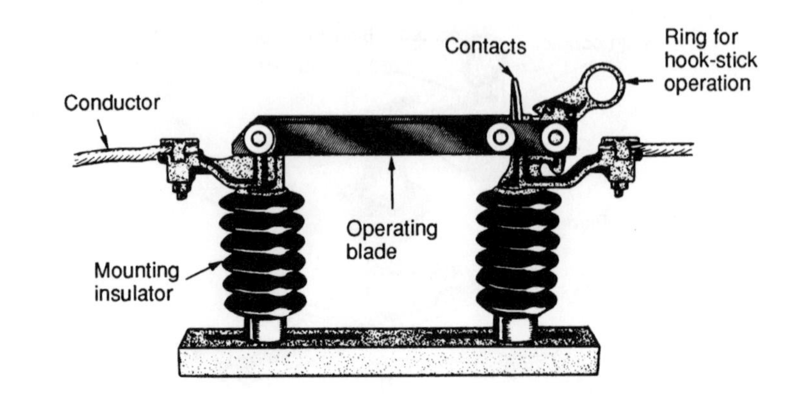

Figure 4-28. A disconnect switch.

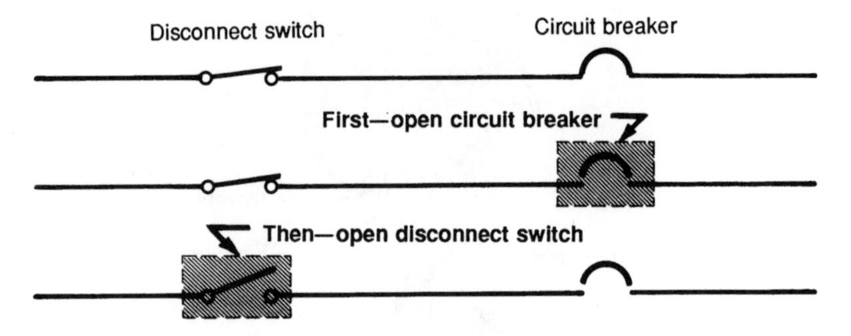

Figure 4-29. The disconnect switch is never opened under load.

action is repetitive. In a recloser, there is an operating rod actuated by a solenoid plunger which opens and closes the contacts, whereas the repeater fuse works only when the metal has been melted by overheat. Figure 4-31 shows a typical single-phase oil-circuit recloser.

However, should it be desirable to delay the action of the recloser, it can be done by an ingenious timing device. Figure 4-32 illustrates an oscillogram showing a typical example of a recloser operation. Notice that the first time it opens and closes, the action is instantaneous requiring only 1.6 cycles. The second time the action is delayed to 2 cycles, the third time to 6, and the fourth time to 5-1/2 cycles. Then the recloser locks itself open and a worker must correct the fault and manually close the mechanism.

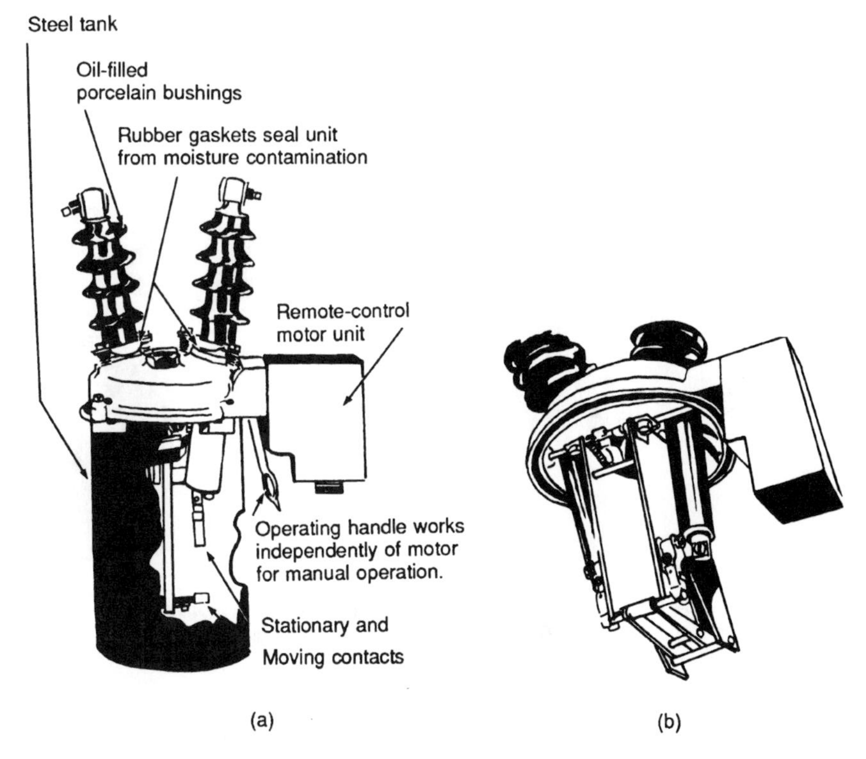

Figure 4-30. An oil switch can be operated manually or by remote control. (a) Cutaway view showing switch in open position. (b) Tank removed showing switch in closed position.

Automated Operation

All of the materials and equipment, assembled properly, make up the primary circuits that serve a particular area. The operation of the circuits include such things as the switching of circuits or portions of circuits (termed sectionalizing) for transferring loads or emergency restoration of service, for controlling voltage regulators, for switching capacitors and other manipulations of lines and equipment-operations usually performed manually under centralized control. Where circumstances generally involving the importance of service continuity, and where justified economically permit, these operations may be accomplished automatically. The relays that actuate these operations may be controlled by preprogrammed computers that impart "instructions" via radio, by telephone lines, or by independent circuits established for that purpose.

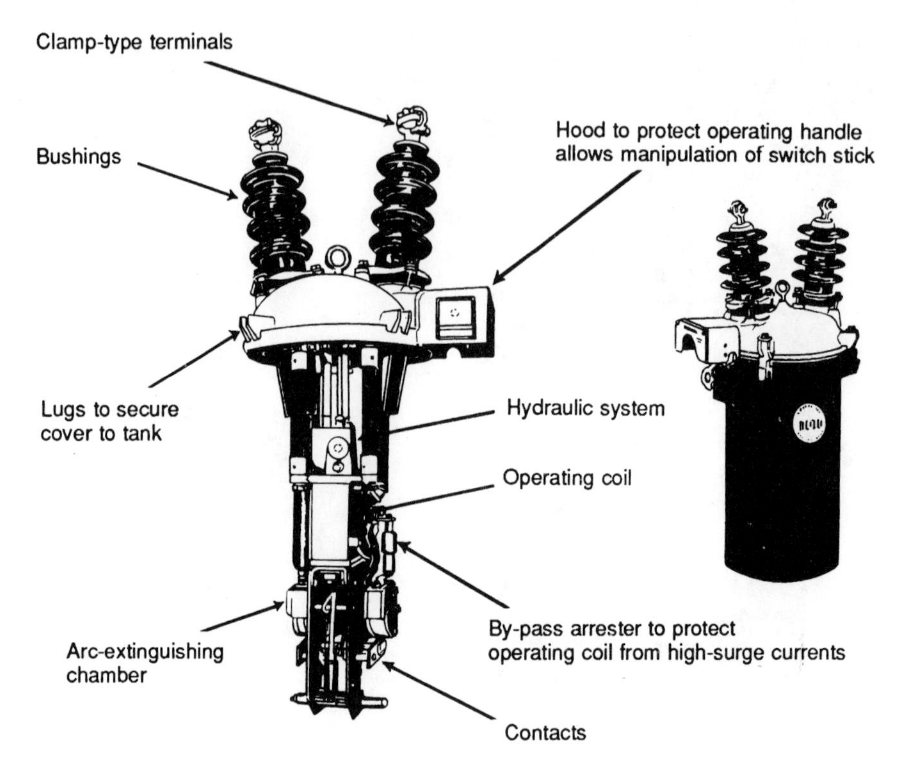

Figure 4-31. Single-phase oil-circuit recloser.

POLYMER (PLASTIC) INSULATION

In general, any form of porcelain insulation may be replaced with polymer. Their electrical characteristics are about the same. Mechanically they are also about the same, except porcelain must be used under compression while polymer may be used both in tension and compression. Porcelain presently has an advantage in shedding rain, snow and ice and in its cleansing effects; polymer may allow dirt, salt and other pollutants to linger on its surface encouraging flashover, something that can be overcome by adding additional polymer surfaces. Polymer's greatest advantage, however, is its comparatively light weight making for labor savings; and essentially no breakage in handling. For detailed comparison of the two insulations, refer to Appendix A, Insulation: Porcelain vs. Polymer.

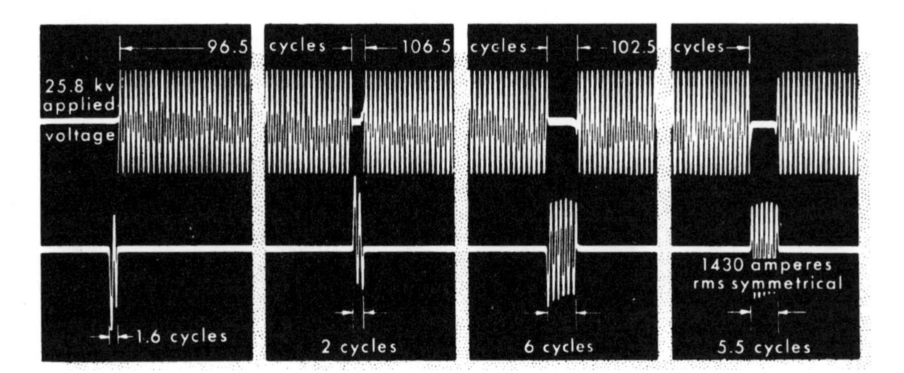

Figure 4-32. Oscillogram showing example of recloser operation.

REVIEW QUESTIONS

1. What is the purpose of a distribution transformer? What are its essential components?

2. Why is the mounting of a distribution transformer important? List some of the methods of doing so.

3. What is the difference between a conventional and a self-protected transformer?

4. What is the function of the fuse cutout?

5. Explain the elementary lightning arrester.

6. List some of the types of lightning arresters and the principles on which they operate.

7. Why is the capacitor so important to the utility company?

8. What are the two broad classifications of switches?

9. What is the advantage of the "load-break" switch?

10. What is an oil-circuit recloser and how does it operate?

Chapter 5
Overhead Construction

OBSTACLES TO OVERHEAD CONSTRUCTION

We have covered some of the basic elements of overhead construction-conductors, insulators, the supports which carry them, transformers, lightning arresters, switches, and fuses. Now consider some of the obstacles that the utility company must overcome when constructing its overhead lines. Trees, pedestrians, railroads, rivers, weather, and hills all present problems which must be overcome. Some of these obstacles are illustrated in Figure 5-1.

CLEARANCES

In constructing overhead distribution lines, the utility company must consider the terrain as well as man-made obstacles such as railroads (see Figure 5-2). For reasons of safety, the voltage of the wires to be strung must be considered. The higher the voltage, the farther away it must be strung from people, traffic, and other wires.

The National Electric Safety Code has set forth specifications governing these clearances. These specifications usually pertain to "class" voltage values. Table 5-1 uses these class values rather than actual voltages in use.

Suppose a 15-kV-distribution line must cross a railroad. According to the specifications given, it must be 30 feet above the ground. In other words, its minimum clearance is 30 feet. But then, a few blocks away, it need only cross a well-traveled street. Here the minimum clearance drops to 22 feet.

Sometimes, wires must cross other wires. Here the voltages of both must be considered to make certain that there is no overstrain on the insulation and no flashover. For instance, if a 750-V wire must cross another 750-V wire, there need only be 2 feet between (see Figure 5-3).

(See also Table 5-2.) However, if one wire should be 110 kV and the other 8.7 kV, requirements call for 6,-1' feet between conductors must be allowed.

Figure 5-1. Obstacles to be overcome in designing and constructing an overhead distribution system.

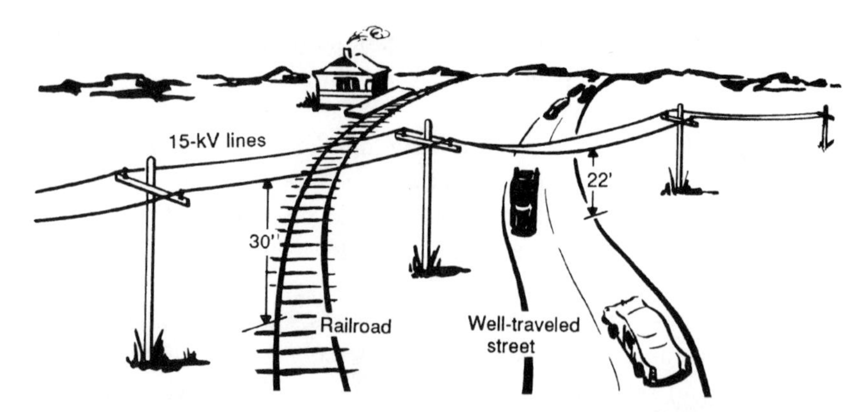

Figure 5-2. Minimum clearances above ground.

Table 5-1. Minimum Wire Clearances Above Ground or Rails (in feet)*

Types of Location	*Guys, Messengers, etc.*	*Voltages*		
		0 to 750	*750 to 15,000*	*15,000 to 50,000*
When crossing above				
railroads	27	27	28	30
streets, roads, alleys	18	18	20	22
private driveways	10	10	20	22
pedestrian walks	15	15	15	17
When wires are along				
streets or alleys	18	18	20	22
roads (rural)	15	15	18	20

*Always refer to the latest edition of the National Electric Safety Code (NESC).

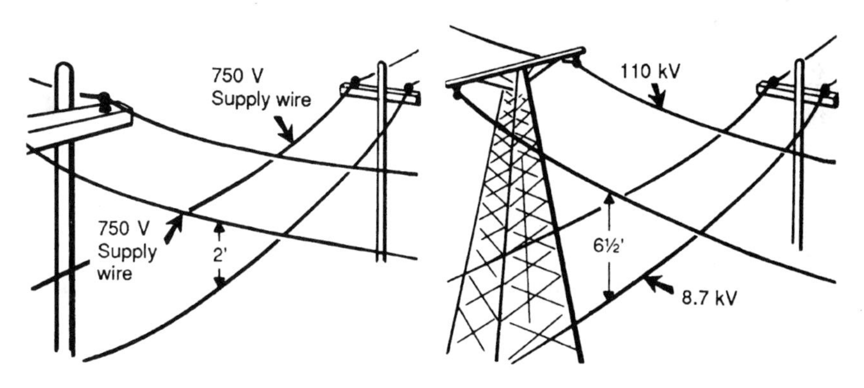

Figure 5-3. Minimum clearances between wires.

Tree Wires

Trees are a frequently encountered obstacle to distribution overhead wiring. Should a tree come in contact with a primary line, it could act as a ground and short the whole circuit. The abrasion of tree branches can often damage the conductor insulation.

Where possible, trees are trimmed. Figure 5-4 shows proper way to trim a tree. Permission for trimming must always be obtained either

Table 5-2. Minimum Clearances Above Other Wires (in feet)*

Nature of Wires Crossed Over	*Commu-nication, Guys*	*Services, Guys, Arrester Grounds*	*Voltages Between Wires*		
			0 to 750	*750 to 8700*	*8700 to 50,000*
Communication circuits	2	2	4	4	6
Aerial supply cables	4	2	2	2	4
Open supply wire					
0 to 750 V	4	2	2	2	4
750 to 8700 V	4	4	2	2	4
8700 to 50,000 V	6	4	4	4	4
Services, guys, arrester grounds	2	2	2	4	4

*Always refer to the latest edition of the National Electric Safety Code (NESC).

from the owner or from local or state authorities. But permission is not always granted; so a tree wire is necessary.

This wire has a tough, outer covering which can withstand considerable abrasion. Figure 5-5 shows a plastic-covered wire which is often used for this purpose. This wire provides partial insulation as well as a tough covering for wires operating at higher voltage values. There are also other types of tree wire.

Plastic coverings have come to be widely used in the power utility industry because they are highly resistant to the action of moisture, minerals, oils, and numerous organic solvents. It must be remembered that such coverings are not to be considered as sufficient insulation for the voltage at which the conductor is operating. The conductor must be handled as if it were bare. Refer to conductors described in Chapter 3.

SAG

Notice how a conductor is strung between poles. It is not pulled tight; it sags. In hot weather it sags even more than in cold weather.

Consider the analogy of the clothesline. A tightly strung clothesline

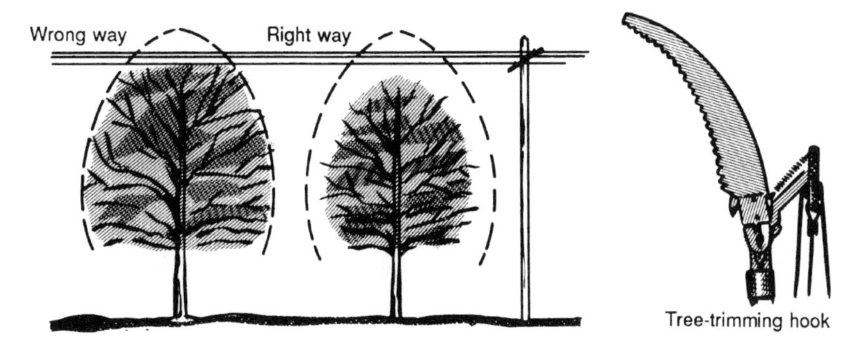

Figure 5-4. Where possible, trees are trimmed.

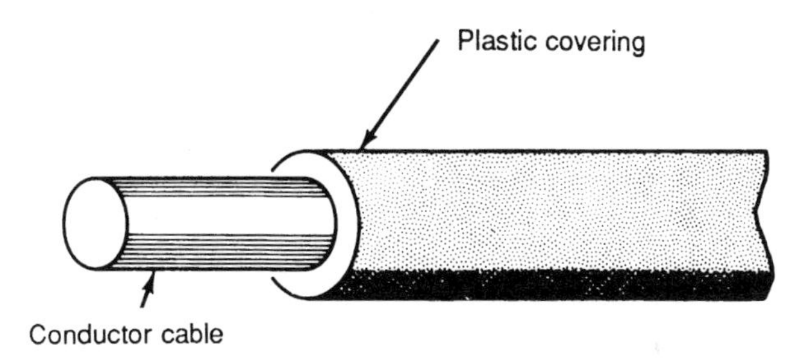

Figure 5-5. Tree wire.

as shown in Figure 5-6(a) puts a much greater strain on the poles which hold it at either end. Adding the weight of wet clothes to a tightly strung line might even pull the poles out of place. However, using a little more rope and allowing the clothesline to dip a little [see Figure 5-6(b)] would relieve this pull considerably. This dip is what is called sag.

The same applies to stringing wire. When metallic wire is strung tightly, it produces a greater strain on the insulator pins and on the pole. Scientific sag is an important factor in stringing wire.

How the Weather Affects Conductors

As previously mentioned, a line sags more in hot weather and less in cold weather. The reason for this is because conductors expand in hot weather; in other words the length of the conductor increases as the temperature increases. It follows that in cold weather the metallic conductor will be shorter than in warm weather. If the wire were strung without sag, it would snap during cold weather. On the other hand, there is a chance a wire strung with too much sag would dip below the specified clearance in warm weather (Figure 5-7).

For copper conductors, the change in length within a temperature range of 100' to 0° is over 5 feet per 1000 feet; for aluminum the change is almost 7 feet.

Besides the temperature, there are other factors that must be considered in determining the sag of a conductor; for example, the length of the span, the weight of the conductors, wind, and ice loading. Also, conductors must not touch each other when swaying in the wind. Tables have been drawn up which consider all these factors and help the utility company determine proper sag.

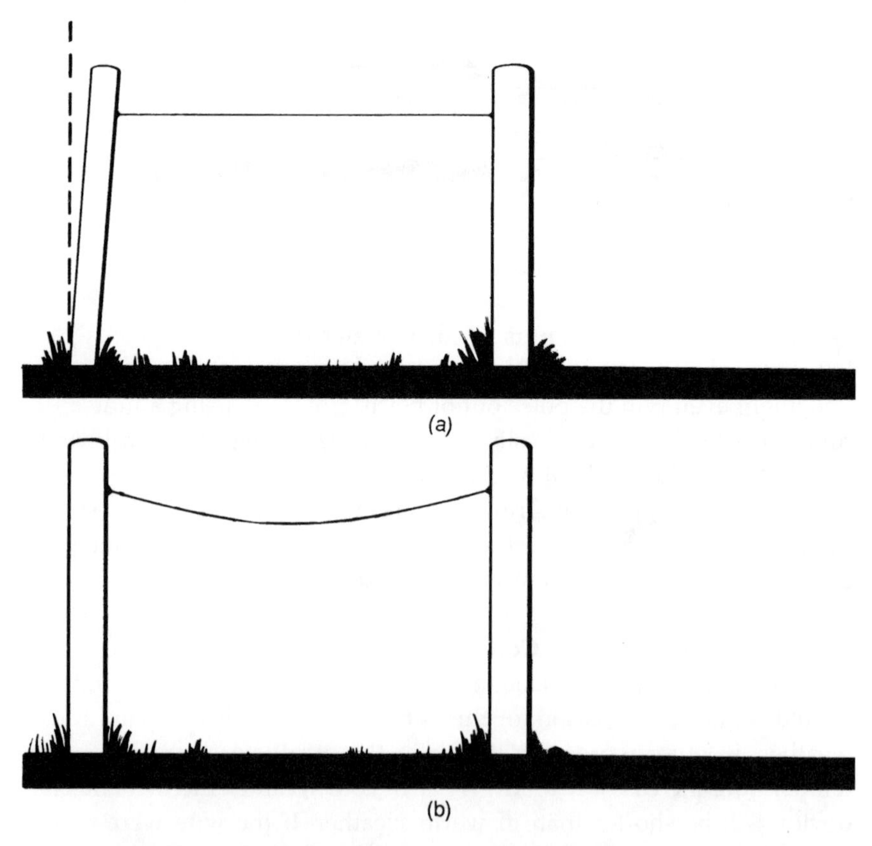

Figure 5-6. Illustration of sag. (a) A taut line causes strain on poles. (b) With a properly sagged line, poles can easily withstand strain.

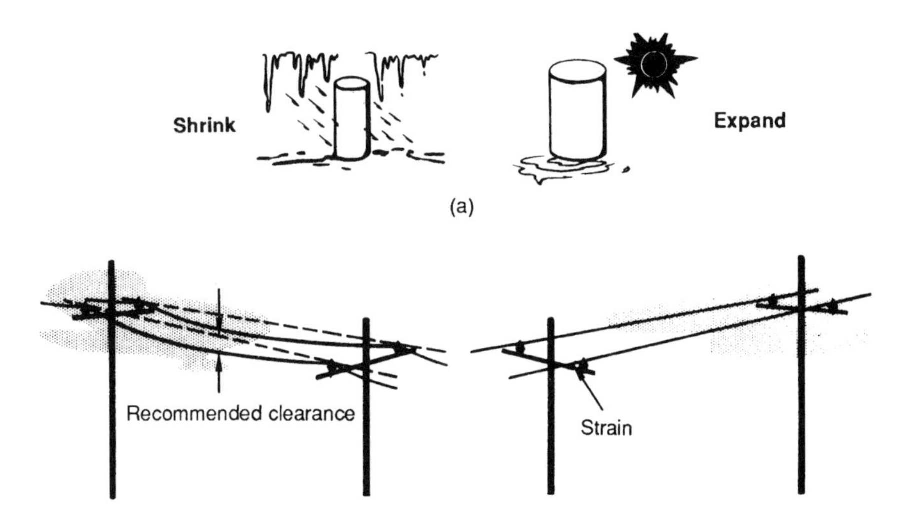

Figure 5-7. Conductor expansion. (a) Conductors shrink in cold weather and expand in hot weather. (b) The line must never sag below the minimum clearance. (c) The line must never be so tight that it strains the poles.

GUYING POLES

Guys

Earlier great detail was given about how carefully poles are chosen to carry the load placed on them by the conductors. Careful specifications have been drawn regarding the length, strength, measurements, and setting depth of a pole for every individual situation. In spite of all this care and planning, situations arise where the conductor tries to force the pole from its normal position. This happens because of abnormal loads of ice, sleet, snow, and wind as well as because of broken lines, uneven spans, corners, dead-ends, and hills.

When these cases of strain arise, the pole is strengthened and kept in position by guys. Guys are braces or cables fastened to the pole. The most commonly used form of guying is the anchor guy (see Figure 5-8).

Anchor Guys

The major components of an anchor wire, which are illustrated in Figure 5-9, are the wire, clamps, the anchor, and sometimes a strain insulator. The wire is usually copperweld, galvanized or bethanized

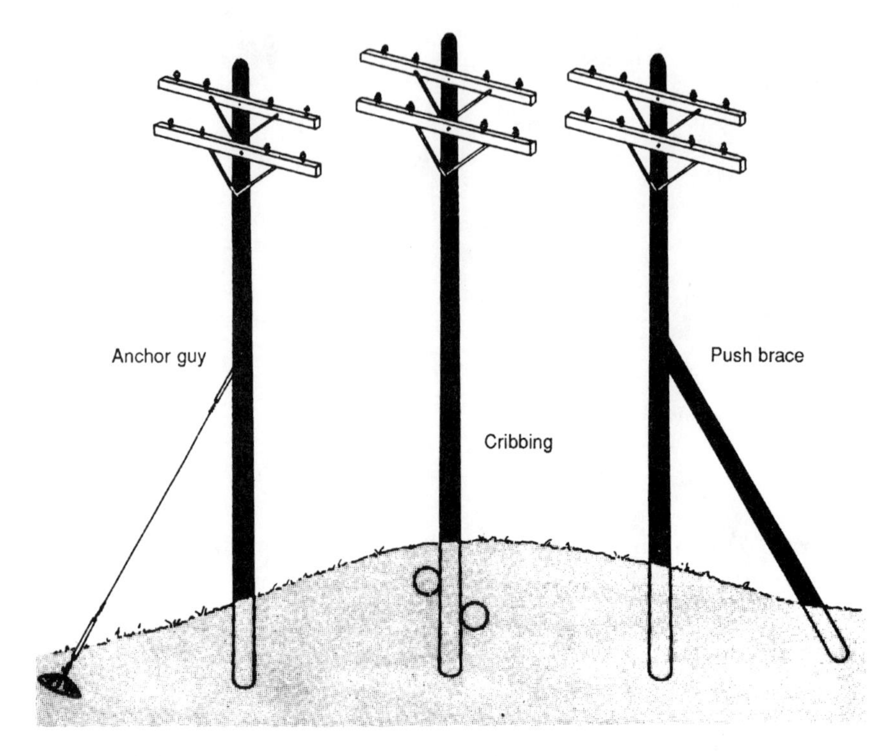

Figure 5-8. Guys are used to keep poles in position.

steel. The guy is usually firmly attached to the pole by a thimble-eye or by a guy eye bolt and a stubbing washer.

The wire is held firmly by the clamp while the bolt is being tightened. Strain insulators were once commonly used on guy wires; however, they are unnecessary in grounded systems (which are more prevalent).

A guy rod forms a connection between the anchor and the guy wire.

Guy Guards

In the construction of anchor guys, the safety and welfare of the public is of primary concern to the utility company. In well-traveled areas, the part of the guy wire nearest the ground is covered with a protector or guard. This serves a three-way purpose. (1) The guard makes the wire more visible to prevent pedestrians from tripping; (2) should a person walk into a guy wire, there is no chance of being cut by

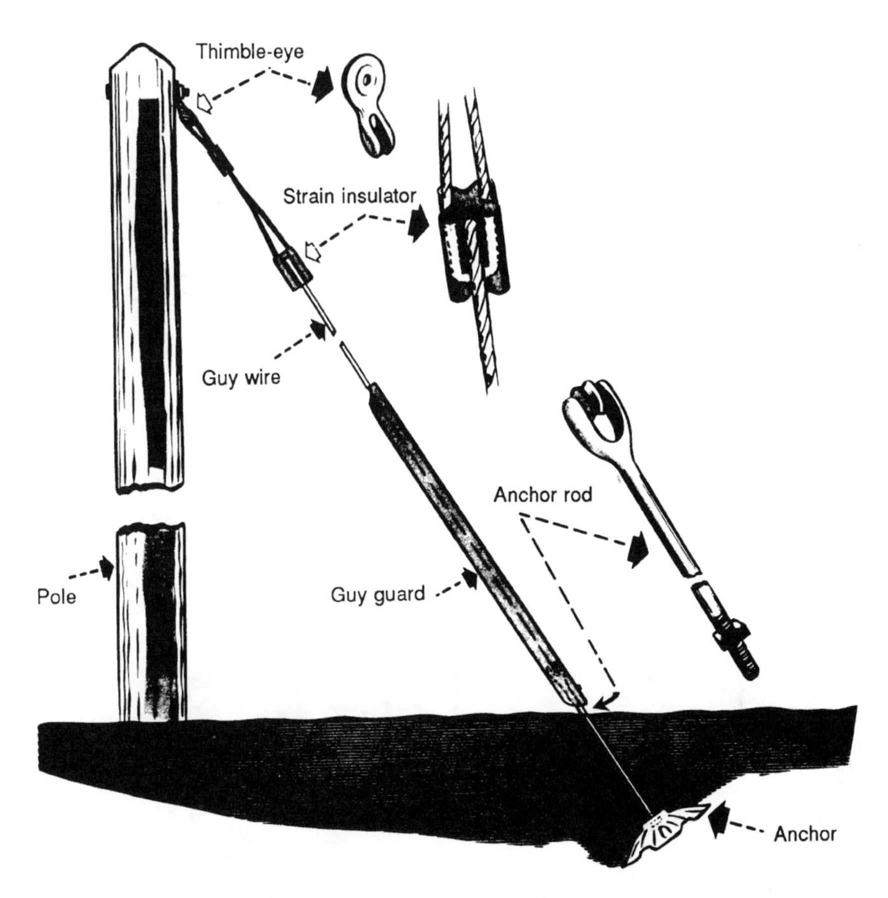

Figure 5-9. Components of an anchor-guy installation.

the wire; and (3) it protects the guy wire from damage by cars or vandals.

At one time a guy guard consisted chiefly of a round metal tube or some wooden blocks. The semiopen guard shown in Figure 5-10 was later developed with rounded edges. This design is cheaper and easier to install than other types of guy guards.

Sometimes temporary guying is necessary. In other words, there is a chance that the land where the anchor is planted may later become "off-limits" to the utility company. For example, a building may be built there, or the owner may decide to construct a driveway. Whatever the reason, screw type anchors (see Figure 5-11) are generally used in temporary situations because they are easily retrievable. They need only be screwed back up and they are ready to be used again. Extra large screw anchors are also used where the soil is swampy or sandy.

Anchors

The value of an anchor is determined by its ability to hold the guy wire under strain. At one time, logs (called dead-men) were buried in the ground to anchor the guy wire. Initially, this was a pretty solid anchor. However, soil conditions often deteriorated the wood. In any case, the digging of a hole large enough to bury a log is inordinately expensive.

Today, manufacturers offer a wide variety of anchors-one to suit every type of ground and every particular situation. Anchors can be installed securely in the ground from swamp to solid rock (see Figure 5-12).

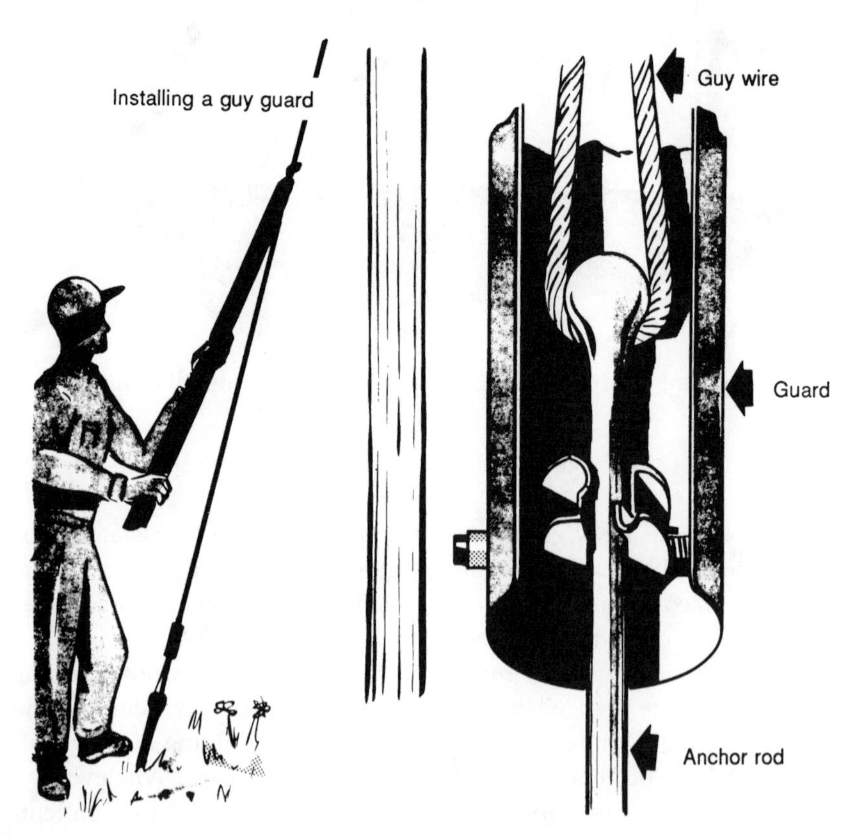

Figure 5-10. Guy guards.

Types of Guys

In situations where an anchor guy is impractical because of crossing a road, a span guy is used. In this form of guying, the guy wire extends from the top of the pole to be guyed to the top of another pole across the street, at approximately the same height. A span guy merely transfers some strain from one pole to another [see Figure 5-13(a)]. Hence, a head guy or anchor guy [see Figure 5-13(b)] is usually used with a span guy.

The head guy is a variation of the span guy, differing in that the wire runs to a point somewhere below the top of the sustaining pole. This type is rarely used in crossing a highway.

Specifications have been set up regarding clearances for guys, just as for conductors.

When it would be difficult to achieve these clearances, a stub guy is used [see Figures 5-13(c) and 5-13(d)]. A stub is just a piece of wood to which the guy wire is attached. The guy must be attached to the stub at some point 8 feet or more above ground. The stub guy wire must allow enough clearance for traffic.

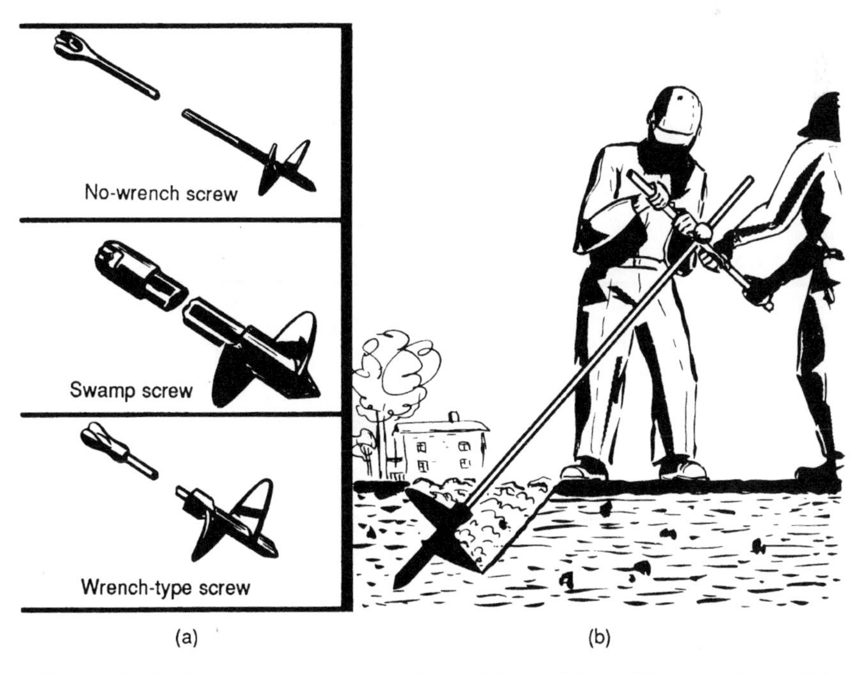

Figure 5-11. Screw-type guy anchors (a), and installing anchors (b).

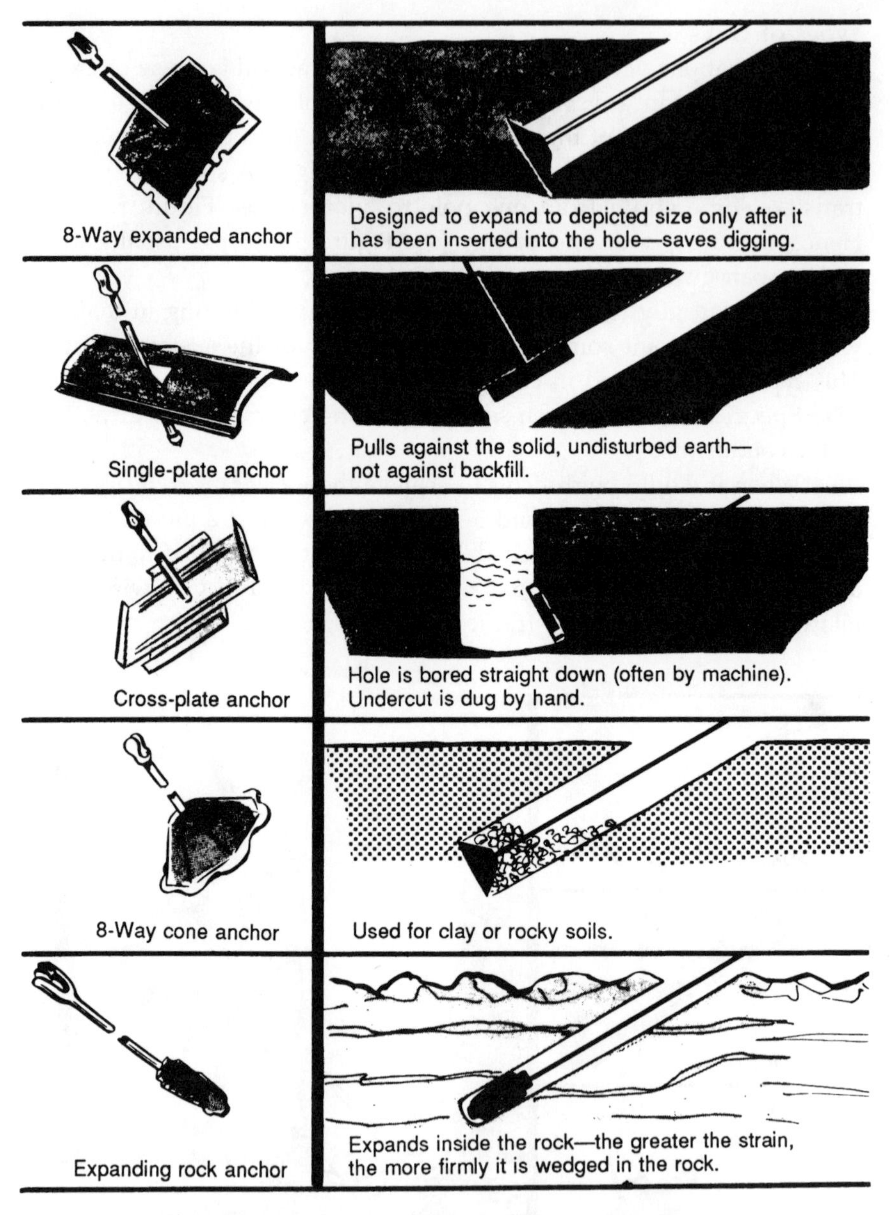

Figure 5-12. Guy anchors and their installation.

Should it prove necessary to guy a pole on private property, some problems may arise. Sometimes owners object to a stub or anchor being planted on their land, but they may approve of a guy wire being attached to a tree [see Figure 5-13(e)]. This is certainly not recommended practice, because a live tree is a very undependable sustainer. However, there are cases where there is no other practical solution.

Where Guys Should Be Used

When a conductor terminates (dead-ends) on a pole, as shown in Figure 5-14(a), a guy is attached to the pole to counteract the pull of the conductors. In case of heavy construction where one guy is inadequate, a guy may be used on the pole next to the last one.

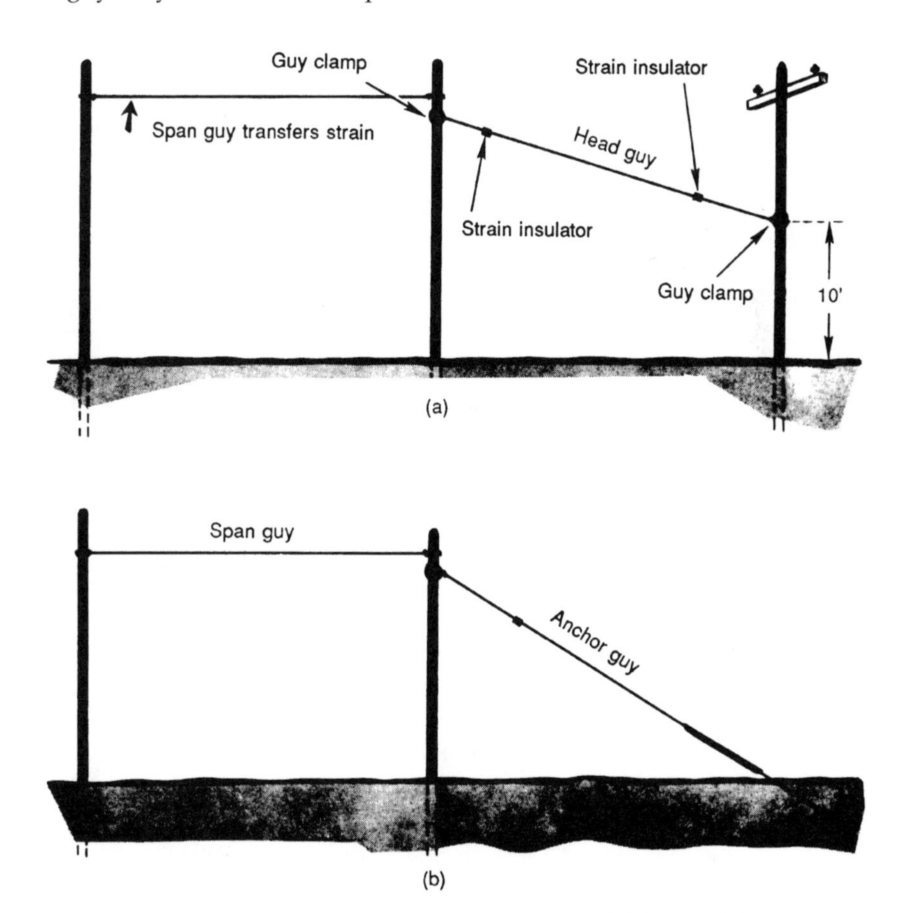

Figure 5-13. How a span guy is used (a) and (b).

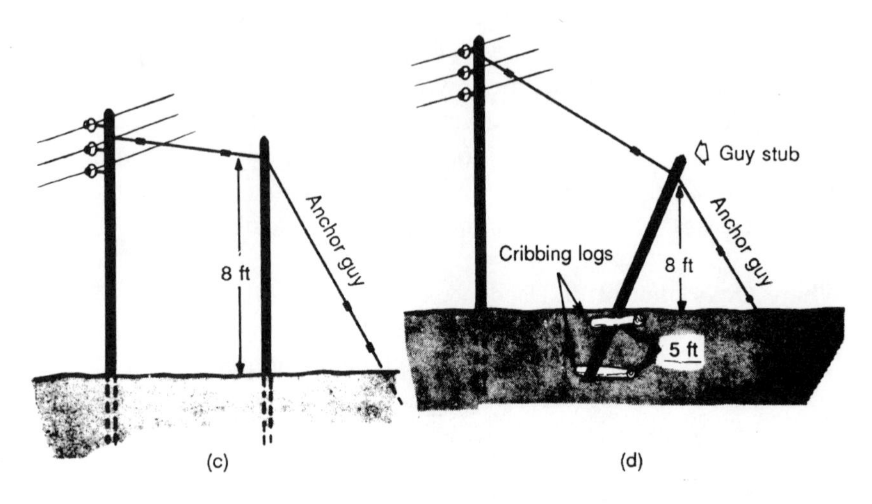

Figure 5-13. (c) Straight and (d) raked stub guy.

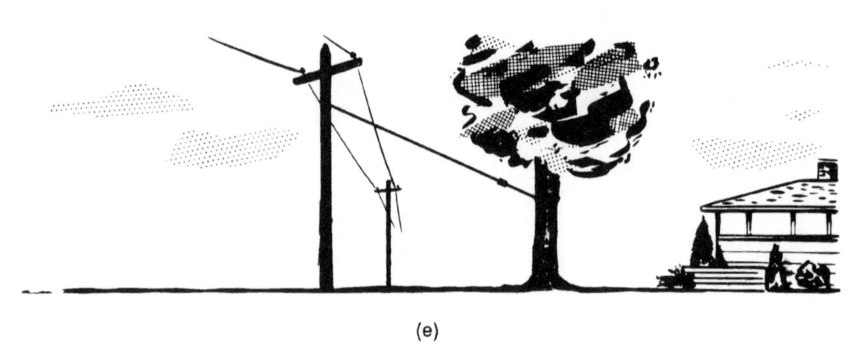

Figure 5-13. (e) Tree used as stub guy. Not recommended practice.

Sometimes there is the problem of a partial dead-end. In other words, more wires are dead-ended on one side of a crossarm than on the other. A guy is run from the side of the crossarm undergoing the worst strain to the adjacent pole [see Figure 5-14(b)].

Corners are treated like dead-ends; however, since the pull on the pole is in two directions, it is wise to use two guys to counteract the pull in two directions, counterbalancing the pull of one set of wires [see Figure 5-14(c)].

When the conductors form an angle, the pole is submitted to an additional stress. To balance these forces, side guys are attached as shown in Figure 5-14(d) to the poles to take up the side pull.

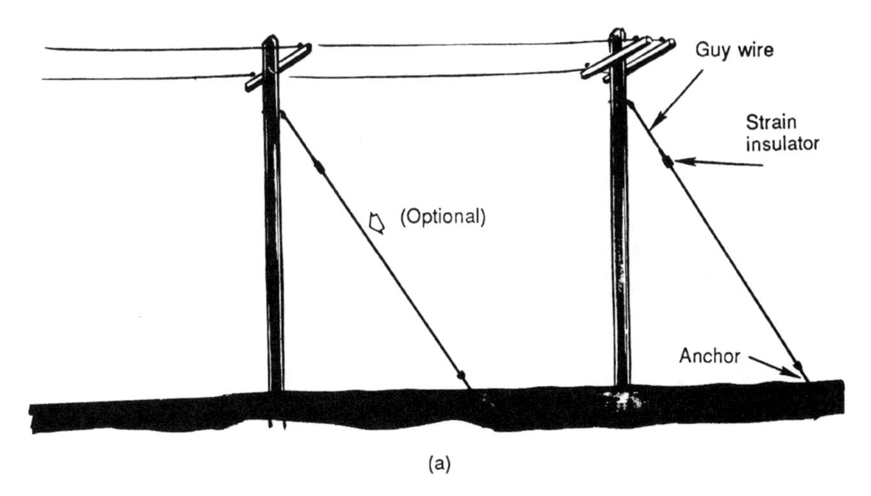

Figure 5-14. (a) Guy installed on terminal or dead-end pole.

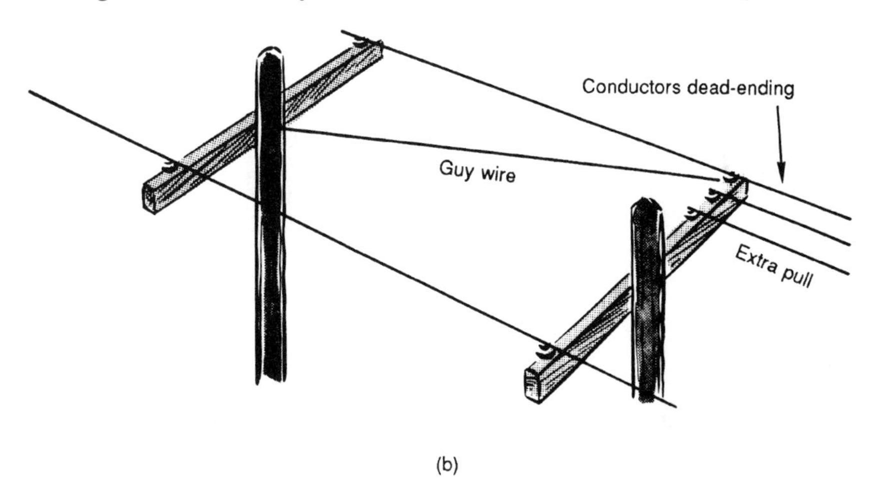

Figure 5-14. (b) Crossarm guy compensates unbalanced pull on crossarm.

When a pole is located on the slope of a hill, a head guy or an anchor guy can be used to counteract the down-hill pull of conductors [see Figure 5-14(e)].

When a branch line shoots from a pole, the weight of the branch conductors causes a side pull. This side pull is counterbalanced by side guys [see Figure 5-14(f)].

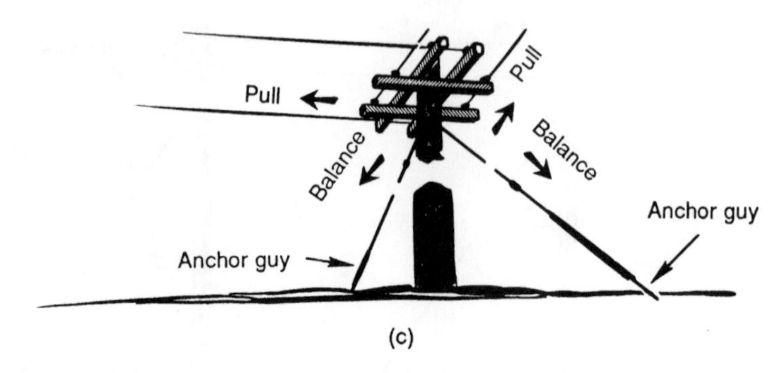

Figure 5-14. (c) Two anchor guys sustaining a corner pole.

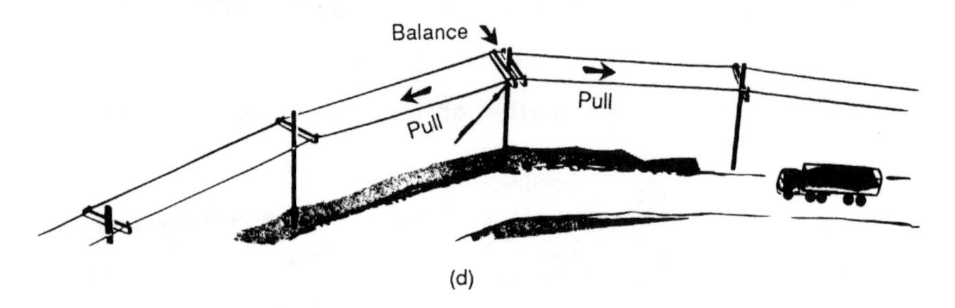

Figure 5-14. (d) Line making an angle.

When a line may be subjected to storms or other strong atmospheric disturbances, storm guys are placed at regular intervals in the line. Generally, four guys are used for this purpose-two line guys and two side guys. All four guys are fastened to the same pole [see Figure 5-14(g)]. Line guys are attached from one pole to the other, thus doing away with the necessity for anchors. Should one pole fail, damage will be limited to the relatively small area between adjacent storm-guyed poles, rather than have the entire line fall, pole after pole.

Other Methods of Sustaining Poles

Sometimes it is necessary to set poles in marshy or swampy land. Since the ground does not hold the pole firmly in such a situation, a guy is necessary. However, it is likely that the use of head or anchor guys might be impractical or impossible for one reason or another. For example, the ground might be too marshy to hold any pole, even a stub; or there might be no available space for an anchor.

Here are some possible solutions.

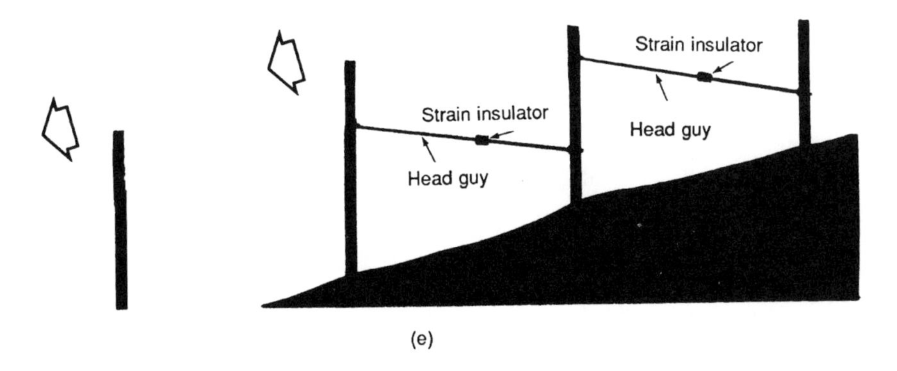

Figure 5-14. (c) Head guys counteract the pull on poles installed on a steep hillside.

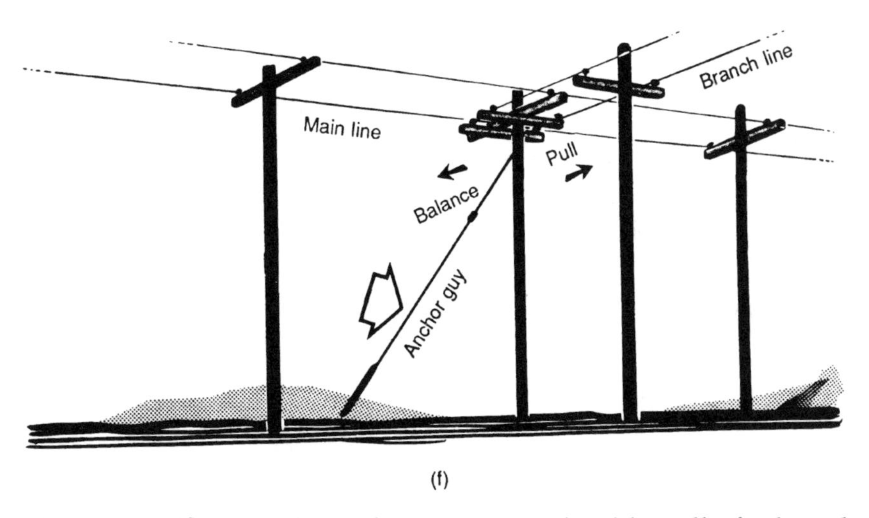

Figure 5-14. (f) A guy is used to counteract the side pull of a branch line connection.

The pole can be propped up with a push-brace [see Figure 5-15(a)].

Before a pole is planted in marshy land, an empty oil drum or a tube of corrugated iron is set in the ground [see Figure 5-15(b)]. After the pole is dropped in, the drum may be backfilled with dirt or concrete.

The area necessary for guying can be minimized by using a sidewalk guy [see Figure 5-15(c)].

Cribbing is necessary in marshy land when the pole must resist an unbalanced load. It is also used in crowded or residential areas when

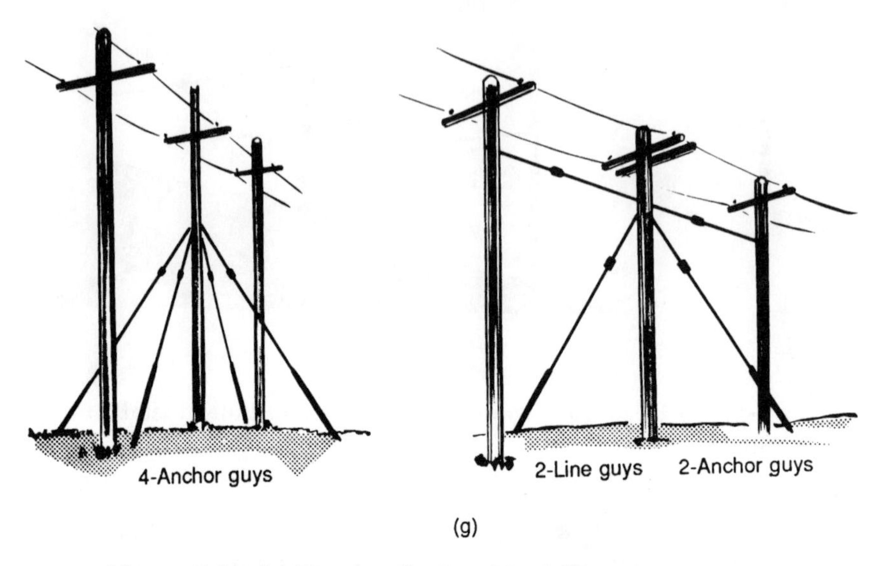

Figure 5-14. (g) Two methods of installing storm guys.

exposed guys would be dangerous, unsightly, or impractical. Cribbing involves placing logs, stones, or other supports at the bottom of the pole on one side and near the surface on the earth on the other side [Figure 5-15(d)). These objects counteract the strain being put on the pole from one direction. As in the case of anchors, utility companies have found it economical in some cases to replace logs with another cribbing device called a pole key [see Figure 5-15(e)].

Joint Construction

For economy and appearance, it is often preferable to use one pole for both power and communication lines (see Figure 5-16). This is known as joint construction. Besides the power companies, telephone companies are the biggest users of poles.

Telephone facilities may consist of open wires on crossarms or of a cable supported on a "messenger." Besides the power and telephone companies, the fire department, the police department, the Coast Guard, the railroad, Western Union, cable TV systems, and even private citizens may have wires strung on the poles.

All these extra wires add a new factor to the problem of line design. Sometimes extra heavy poles or additional guying become necessary, especially in the case of telephone cables. Distance must be

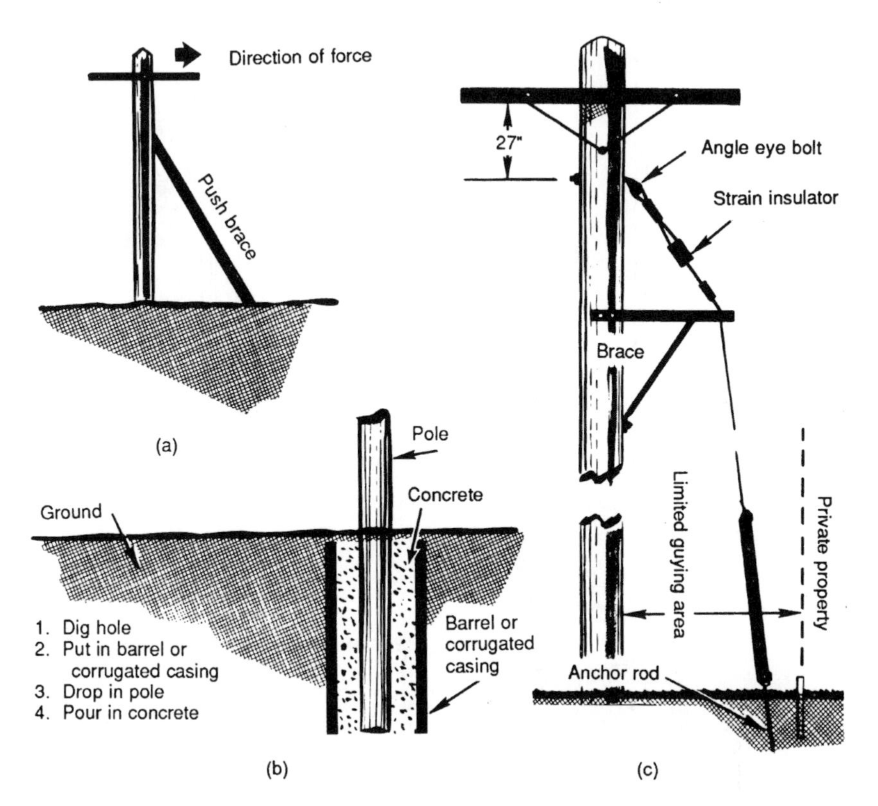

Figure 5-15. Other methods of sustaining poles. (a) Push brace, (b) concrete-filled barrel, and (c) braced anchor buy.

considered carefully between the conductors and equipment of the two power facilities.

Usually the power companies and the communication companies draw up contracts which specify the conditions under which both companies may share the same facilities. Usual specifications include a minimum of 40 inches spacing between these facilities.

OVERHEAD CONSTRUCTION SPECIFICATIONS

Some typical construction standards incorporating previous discussions are shown in Figure 5-17(a) through 5-17(g).

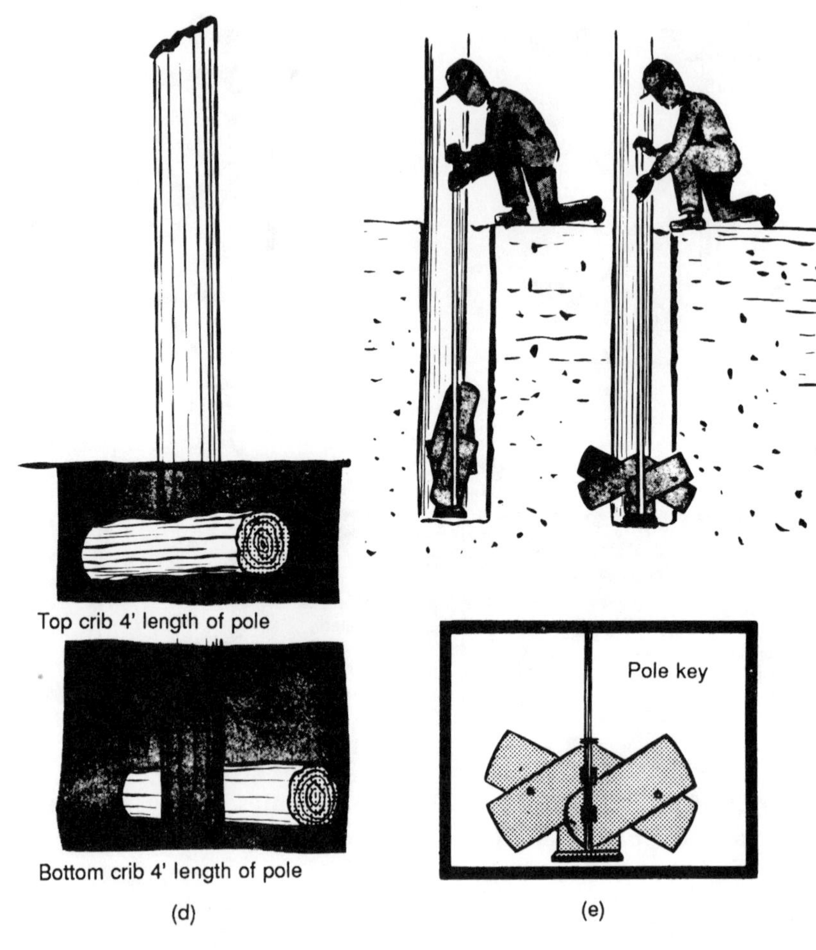

Figure 5-15. (d) Cribbing and (e) installing a pole key.

REVIEW QUESTIONS

1. What factors influence the clearance required for overhead lines?

2. What is meant by the sag of a conductor?

3. What factors affect a conductor's sag?

4. Why is it necessary for conductors to be properly sagged?

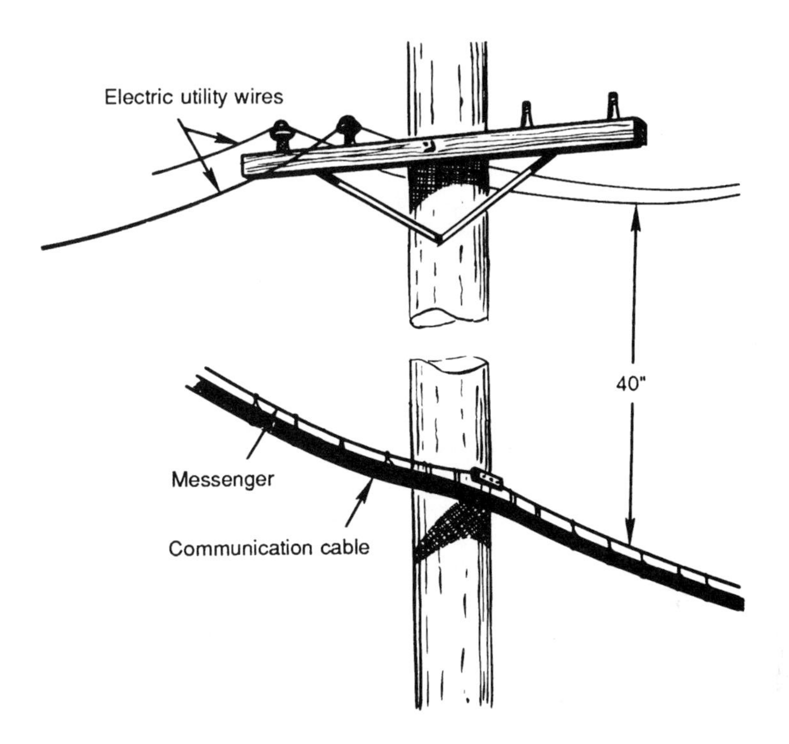

Figure 5-16. Joint construction of telephone and electric cables.

5. What are guys and why are they used?

6. What is the most commonly used form of guying? Name some of its components.

7. Why are guy guards used?

8. List some other ways of sustaining poles.

9. List some situations where guys should be used.

10. What is meant by joint construction?

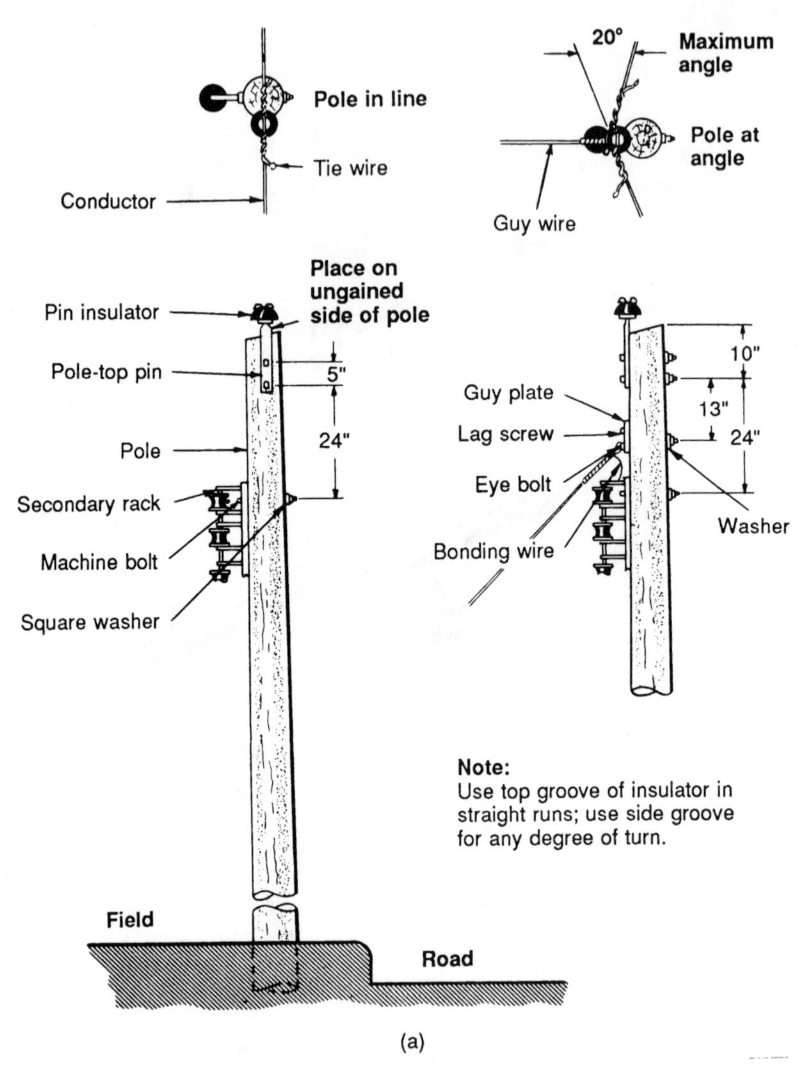

Figure 5-17. Typical construction standards for a primary line assembly. (a) Single-conductor pole head, 0° to 20° angles.

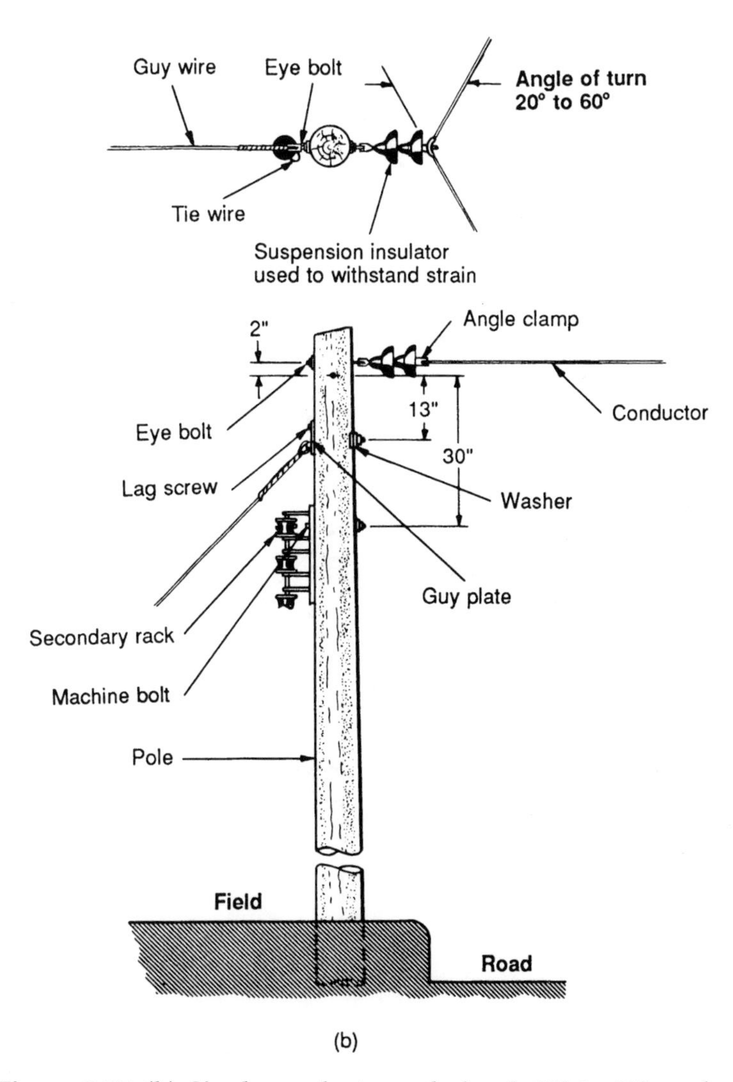

Figure 5-17. (b) Single-conductor pole head, 20° to 60° angles.

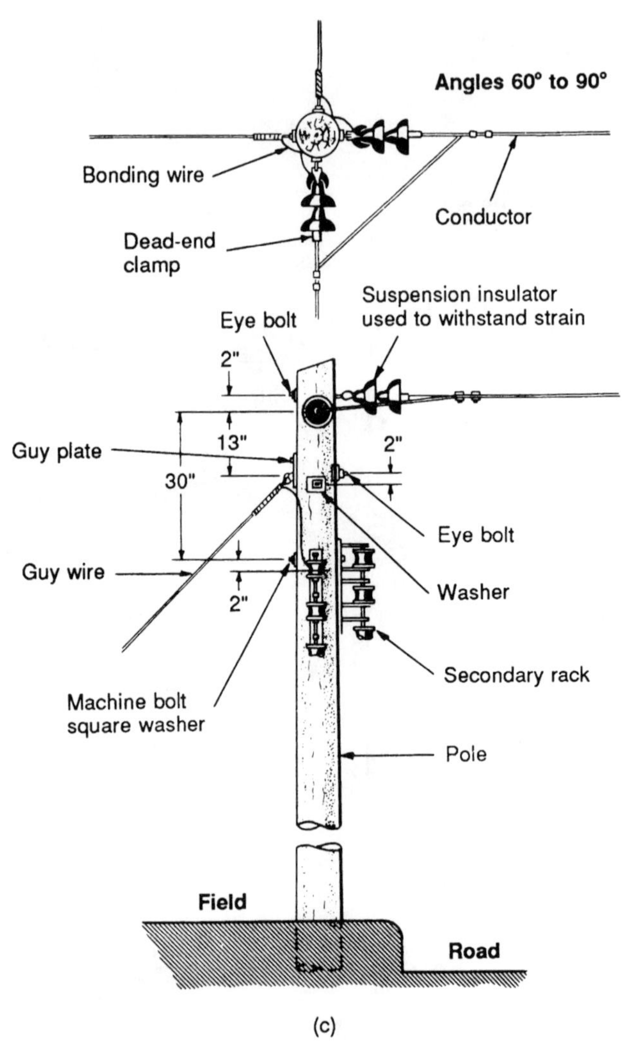

Figure 5-17. (c) Single-conductor pole, 60° to 90° angles.

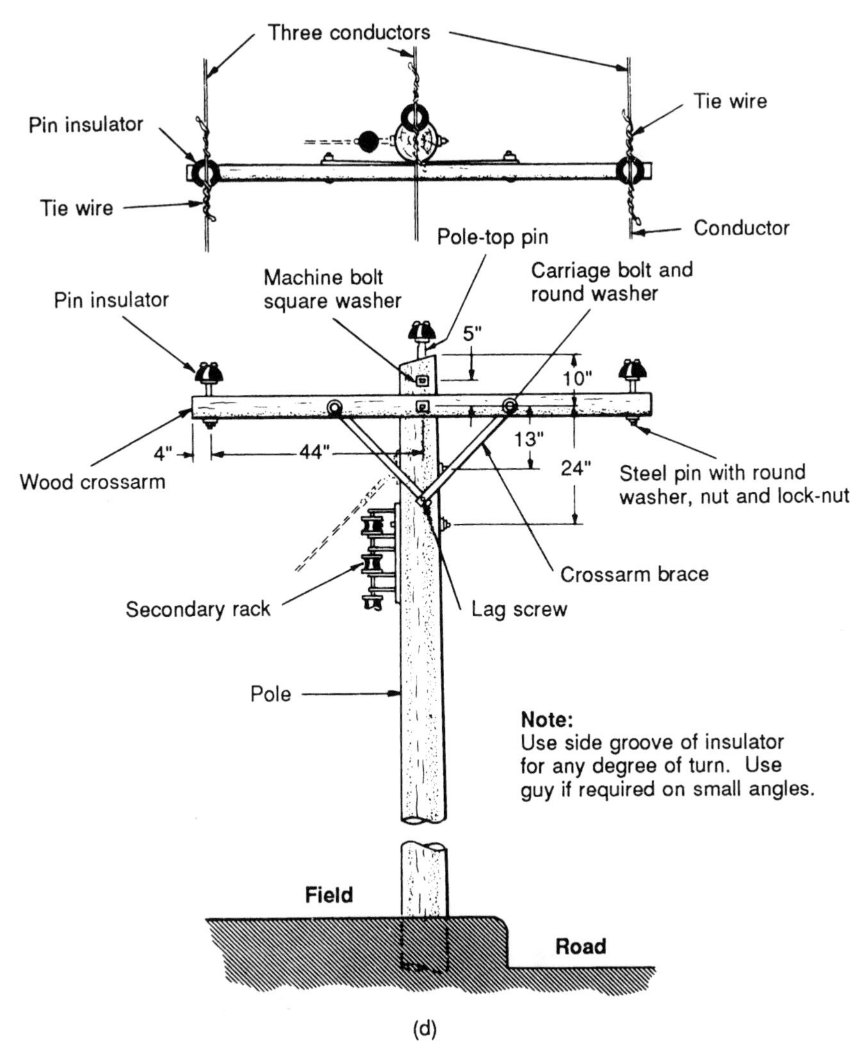

Figure 5-17. (d) Three-conductor pole head, 0° and small angles.

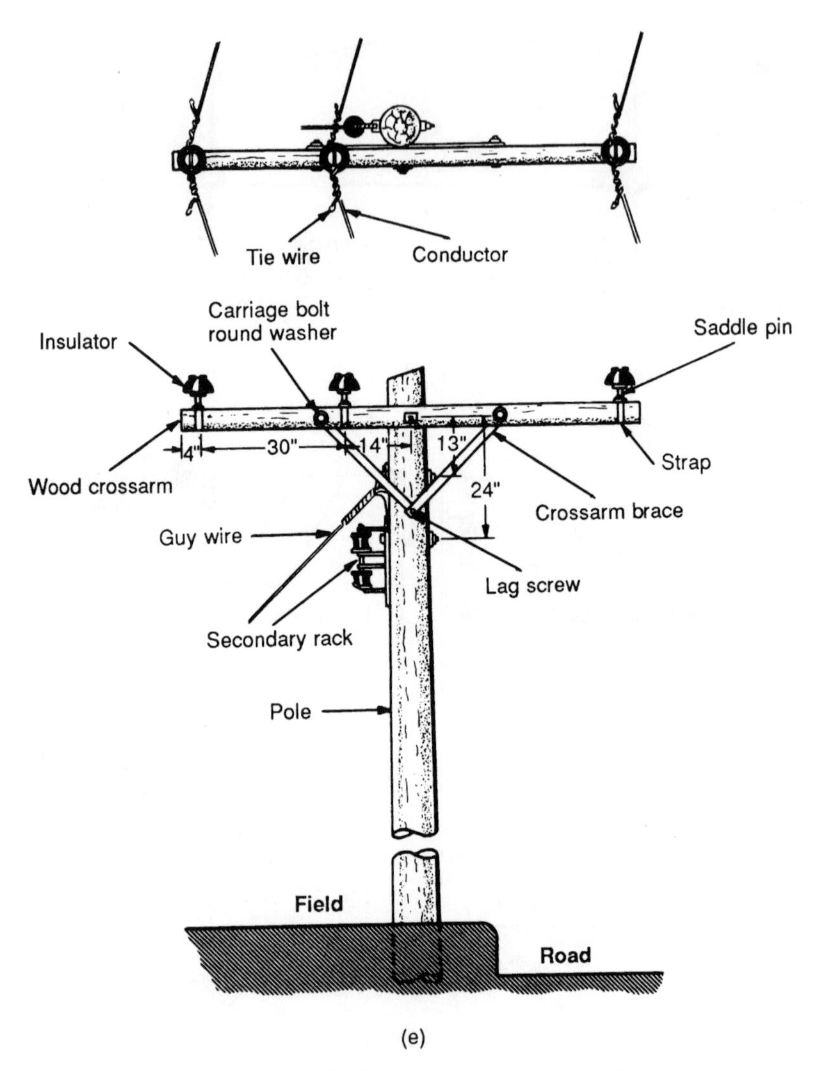

Figure 5-17. (e) Three-conductor pole head, medium angles.

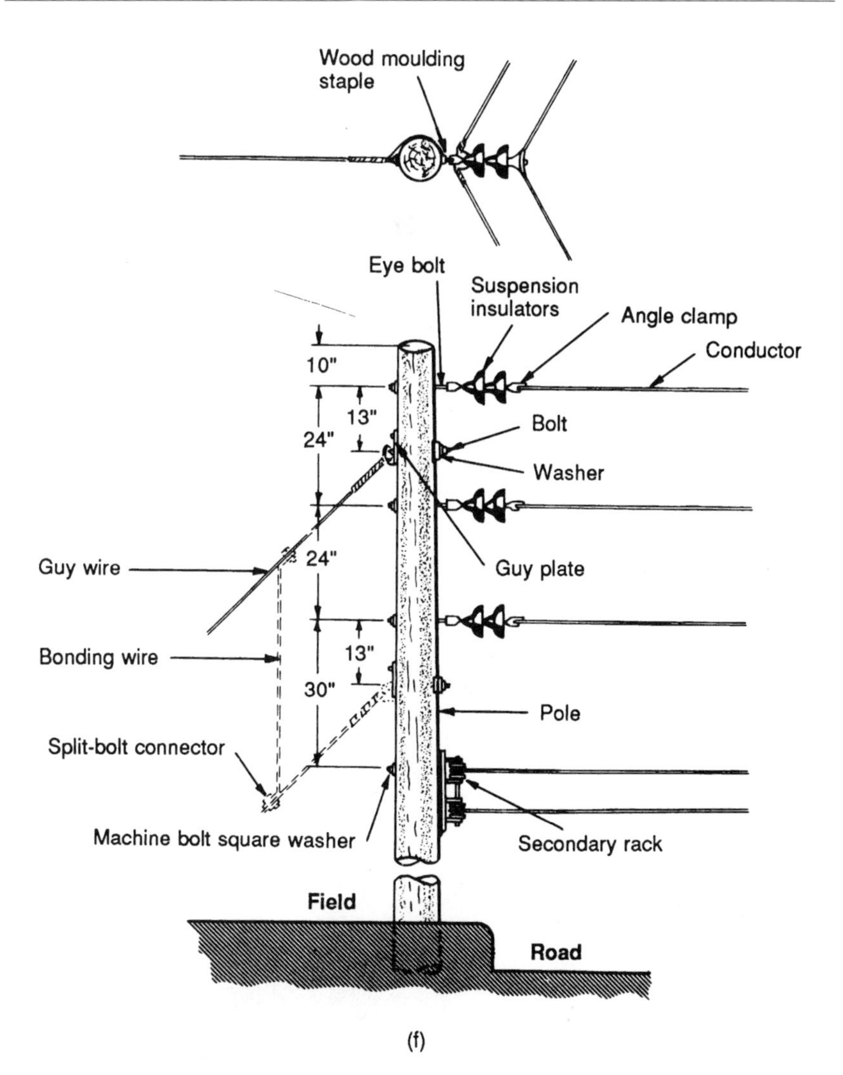

Figure 5-17. (f) Three-conductor pole head, angles up to 60°.

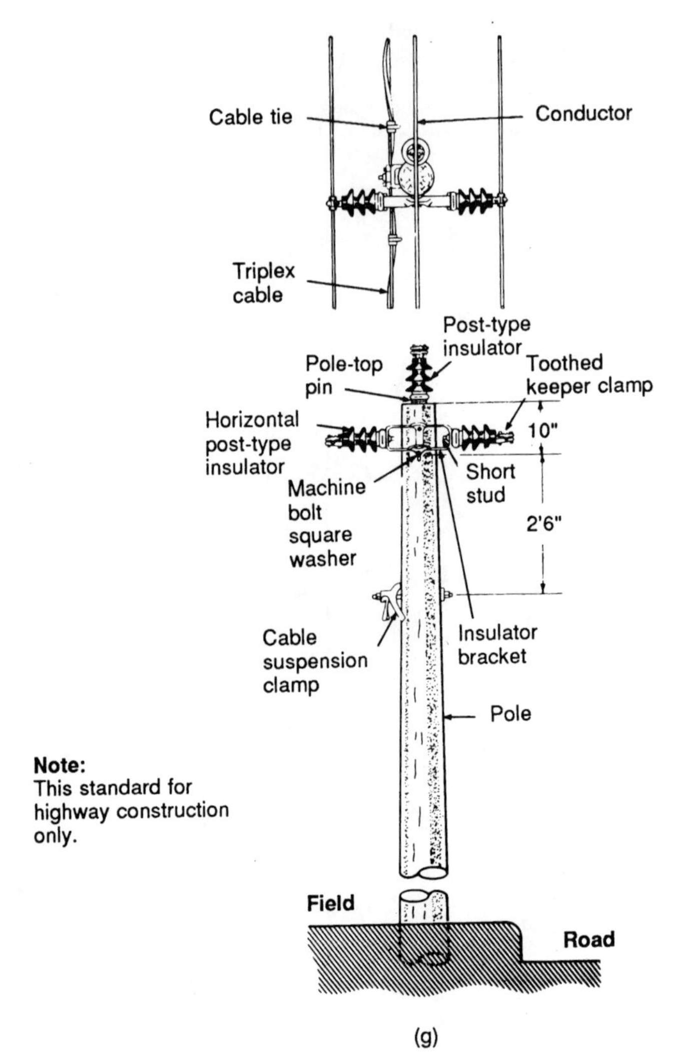

Figure 5-17. (g) Three-conductor pole head, 0° to 15° angles; aimless construction.

Chapter 6

Underground Construction

UNDERGROUND CONSTRUCTION

Where general appearance, economics, congestion, or maintenance conditions make overhead construction inadvisable, underground construction is specified. While overhead lines have been ordinarily considered to be less expensive and easier to maintain, developments in underground cables and construction practices have narrowed the cost gap to the point where such systems are competitive in urban and suburban residential installations, which constitute the bulk of the distribution systems.

The conductors used underground (see Figure 6-1) are insulated for their full length and several of them may be combined under one outer protective covering. The whole assembly is called an *electric cable.* These cables may be buried directly in the ground, or may be installed in ducts buried in the ground. Concrete or metal markers are often installed at intervals to show the location of the cables.

In residential areas, such cables may be buried by themselves by means of a plow or machine digging a narrow furrow. In commercial or other congested areas, where maintenance repair or replacement of the cables may be difficult, conduits or ducts and manholes may be installed underground to contain the cable and other equipment. Figure 6-2 illustrates a typical underground setup of a conduit system.

RESIDENTIAL UNDERGROUND CONSTRUCTION

Underground Residential Distribution Layouts

Distribution circuits to residential areas are similar to overhead designs, except the installation is underground. Primary mains, with

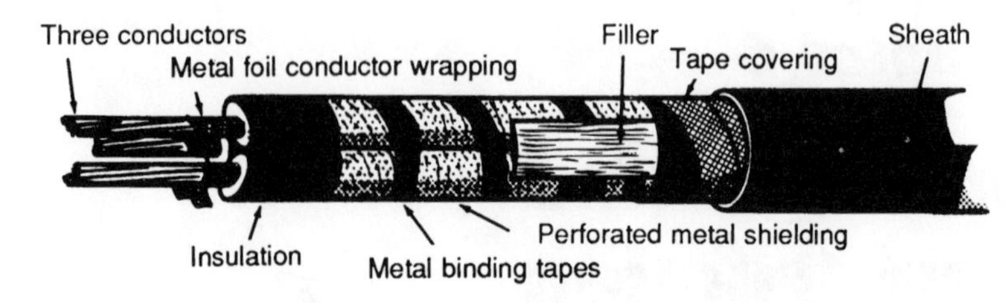

Figure 6-1. An underground cable.

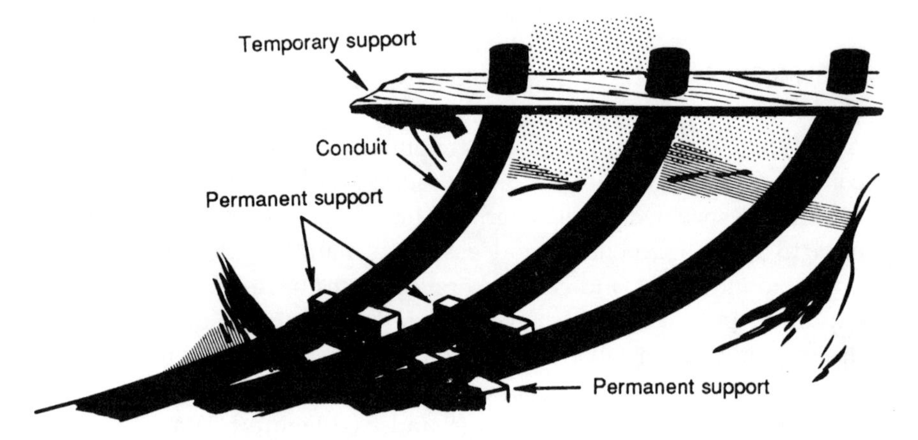

Figure 6-2. Preparing the installation of an underground conduit system.

take-offs, are installed to which are connected the distribution transformers that supply secondary low-voltage (120-240 volt) service to the consumers. Two general patterns have developed, economy being the deciding factor in the selection.

One pattern has as the primary supply a distribution transformer which may feed two or more consumers via secondary mains and services (see Figure 6-3).

Another pattern has as the primary supply individual transformers feeding only one consumer (see Figure 6-4). No secondary mains are required here, and the service connection to the consumer may be practically eliminated by placing the transformer adjacent to the consumer's service equipment.

The primary supply is of the radial type; that is, the feeders supplying a territory "radiate" from their substation source, much as the spokes of a wheel. Similarly, as is done in overhead circuits, the spurs or

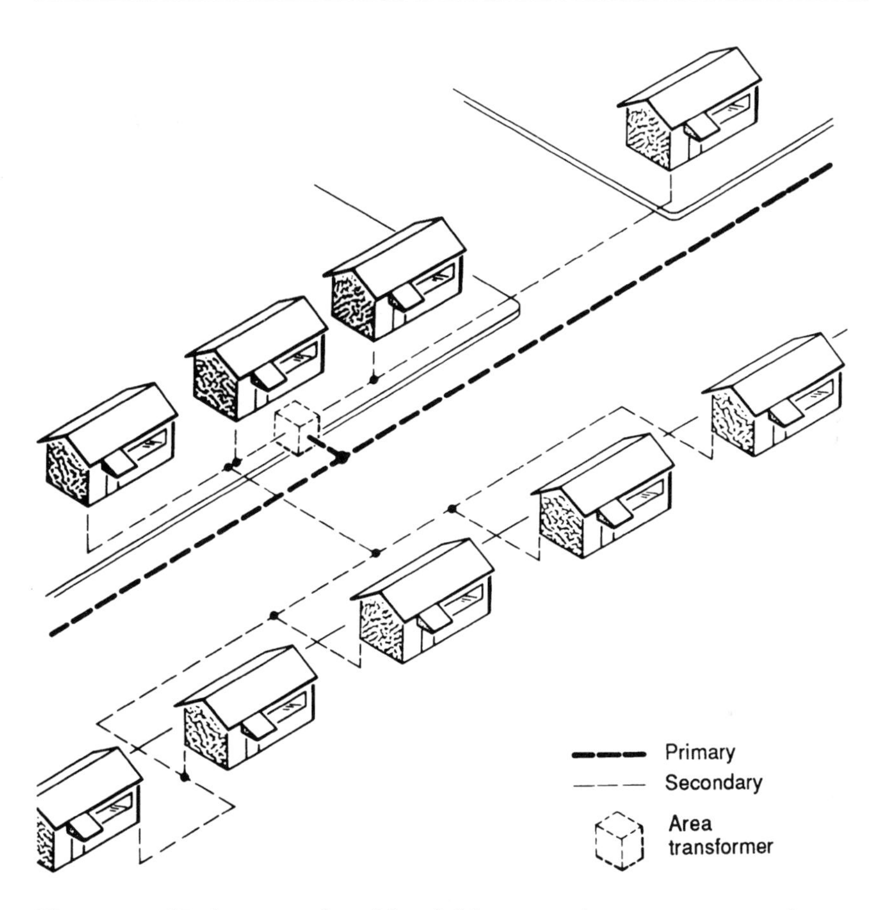

Figure 6-3. Underground residential layout using an area transformer.

laterals are connected to the primary main through fuses, so that a fault on these laterals will not cause an interruption to the entire feeder.

However, as faults on underground systems may be more difficult to locate and take longer to repair than on overhead systems, the primary supplies may often be arranged in the pattern of an open-loop (see Figure 6-5). In this instance, the section of the primary on which the fault has occurred may be disconnected at both ends and service re-established by closing the loop at the point where it is normally left open. Such loops are not normally closed because a fault on a section of the feed may then cause the fuses at both ends to blow, leaving the entire area without supply and no knowledge of where the fault has occurred.

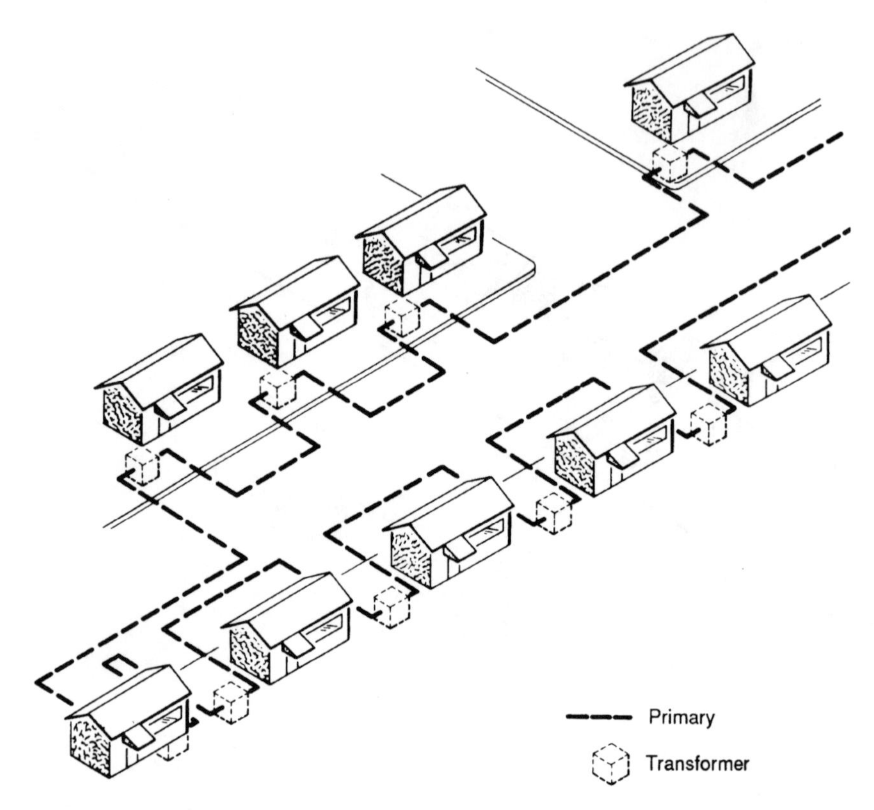

Figure 6-4. Underground residential layout using individual transformers.

COMMERCIAL UNDERGROUND CONSTRUCTION

Underground Commercial and Industrial Supply

Because commercial and industrial loads are heavier than in residential areas, three-phase supply is often installed. That is, in place of a single two-wire—or so-called single-phase—source, a combination circuit of three such phases-or three-phase—as a source is used. Here, the return, or neutral, wire of each of the three phases may be consolidated into one common neutral wire for all of the phases; hence, only four wires are required instead of six, making for economy.

Again, because these consumers may he located in congested areas where digging may be difficult, cables and equipment may be installed in ducts and manholes to facilitate maintenance and replacement.

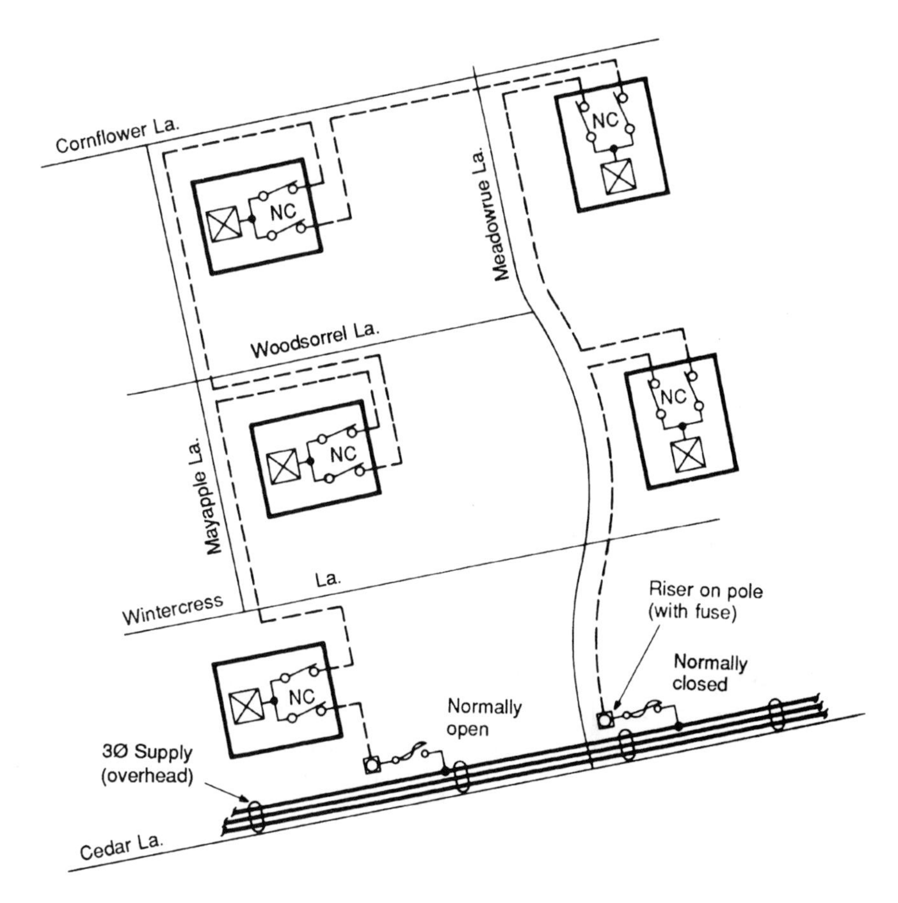

Figure 6-5. Underground residential layout using open-loop construction.

Since some of the operations carried on by consumers of this type are critical, the reliability of electric supply may be of such importance as to warrant extra expenditures for firming up the supply. This may be done by having a second source installed with automatic throw-over devices (see Figure 6-6) so that should one source of supply fail, the service may automatically be switched over to the alternate source.

An even more complex, and expensive, method is to create a low-voltage network by connecting together, through switches called protectors as shown in Figure 6-7, the secondary mains from transformers supplied by several primary feeders. Here, should one feeder become de-energized, the power is still supplied from the others without interruptions. The protector on the de-energized feeder automatically opens to prevent a "feedback" from the other feeders back (through the transformer) to the de-energized feeder.

UNDERGROUND CONSTRUCTION—DIRECT BURIAL

Buried Cable

The installation of cable directly in the ground saves the cost of building conduits and manholes and allows the use of long sections of cable, thereby eliminating the necessity for a number of splices.

The cable may be buried alone in a trench, or may be buried in a trench together with other facilities, including telephone cables, gas mains, water, or sewer systems (see Figure 6-8). Sharing the cost of installing such facilities jointly contributes greatly to the economy of underground distribution systems. The width and depth of the trench are dictated by the number and type of facilities to be installed and by National Electric Safety Code (NESC) requirements; a typical arrangement is shown in Figure 6-9. Where only one cable is involved, to avoid the cost of digging and backfilling the trench, and to disturb the surface as little as possible, the cable is sometimes installed by a plow which furrows the earth to the required depth, lays the cable, and presses the

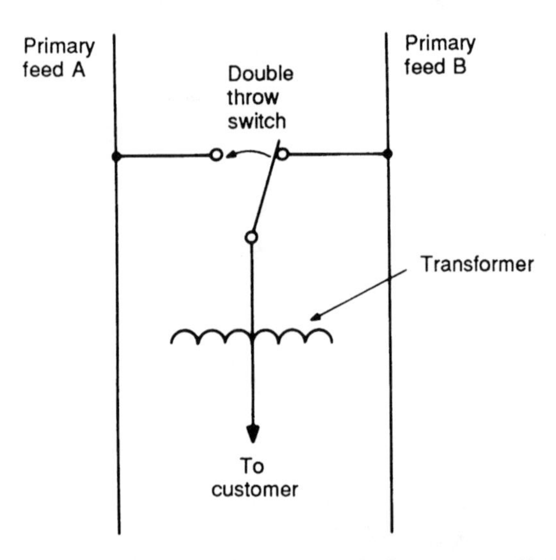

Figure 6-6. Use of throw-over switch for extra reliability.

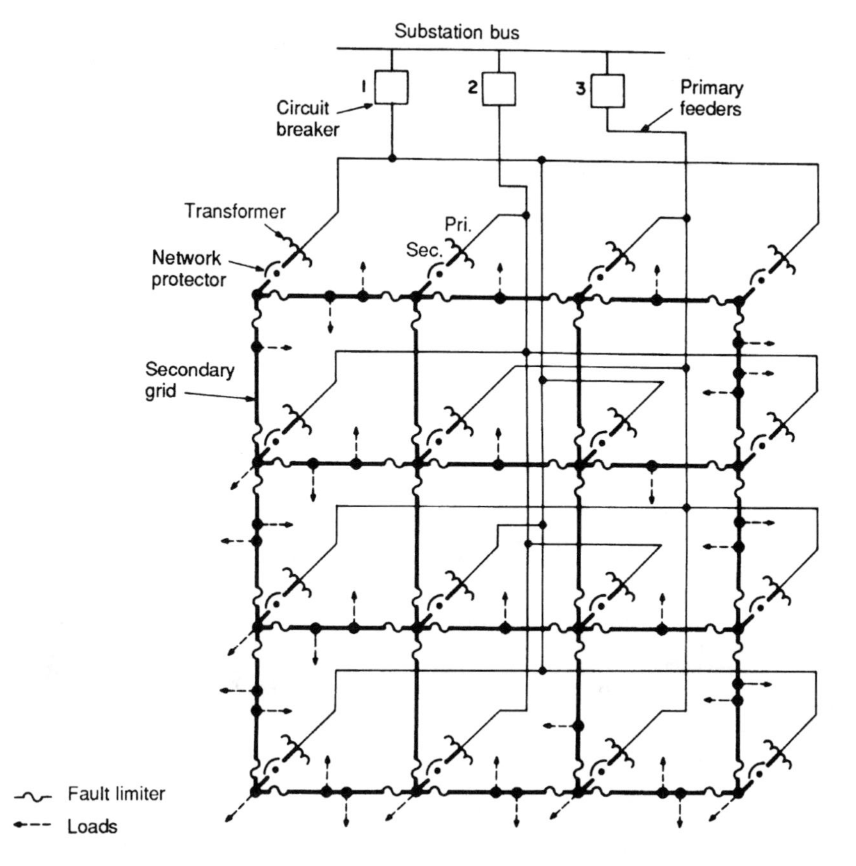

Figure 6-7. Secondary underground network. Protector switches are used to prevent feedback.

sod back in place after the plow has passed. Joints on this underground cable are buried directly in the ground.

UNDERGROUND CONSTRUCTION—CABLES

Underground Cable Sheath

In underground work, cables must be sheathed for protection from mechanical injury during installation, from moisture, gasses, chemicals (see Figure 6-10), and other damaging substances in the soil or atmosphere. Sheathing must have specific properties to meet the varying

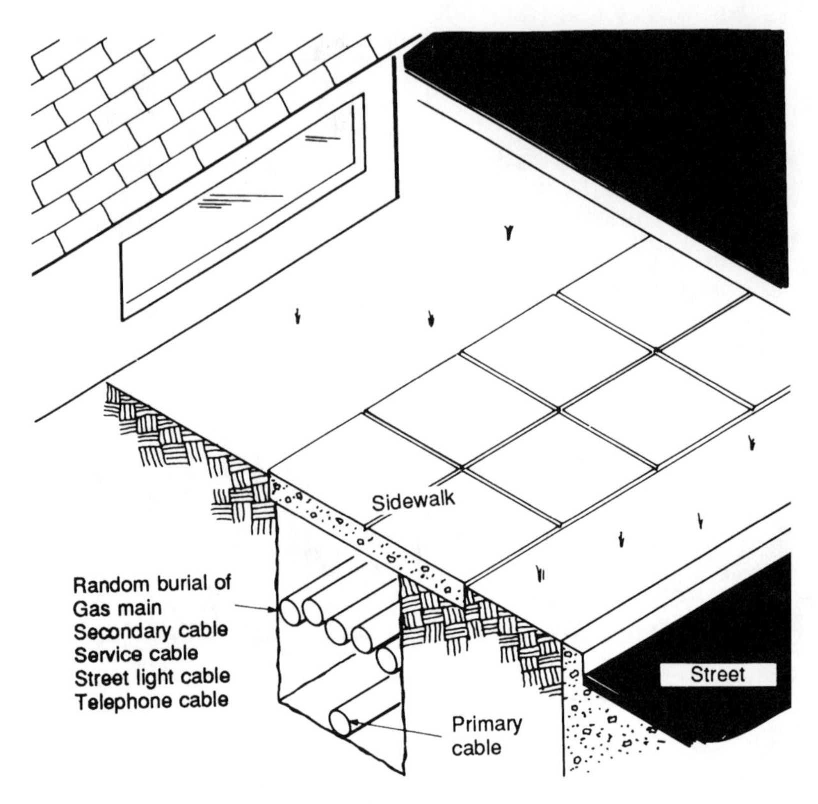

Figure 6-8. Typical "joint" underground distribution system. Random burial of gas main, secondary cable, street light cable, and telephone cable with the primary cable.

conditions. At one time, almost all cables were lead-sheathed because lead is almost completely moisture proof and a good mechanical buffer; but it may be corroded by some chemicals. New types of plastics which do not deteriorate from exposure to any weather, sunlight, moisture, and chemicals have largely taken the place of the more expensive and less flexible lead sheathing. Earlier neoprene, polyethylene, and polyvinylchloride (PVC) sheaths have given way to plastics compounded to meet the more demanding particular needs of direct-burial installations.

The other essential parts of a sheathed cable are the conductors and the insulation.

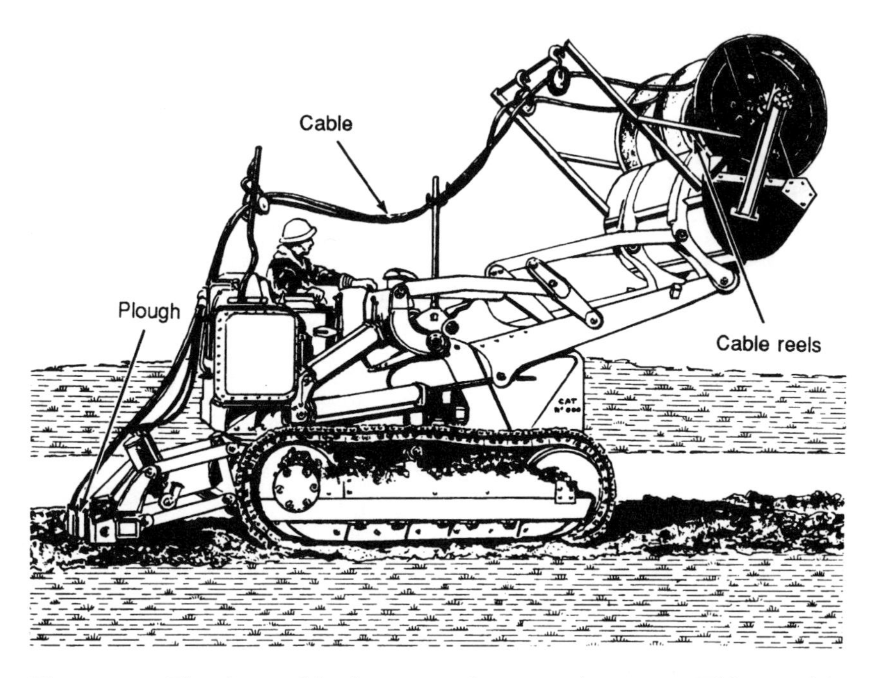

Figure 6-9. Plowing cable for an underground system. This machine digs trench, buries cable, and backfills sod in one operation.

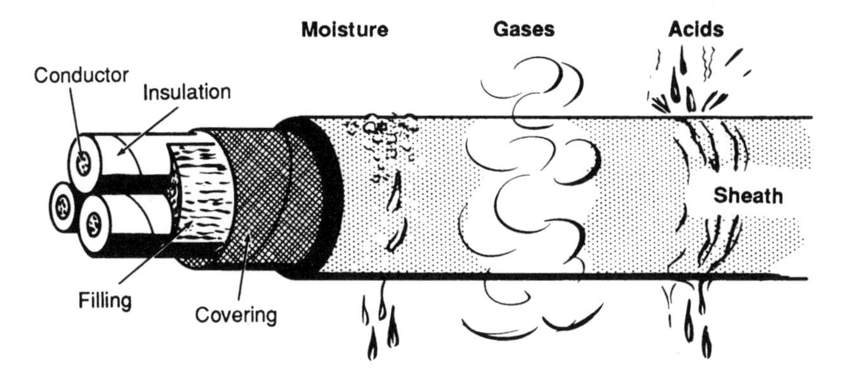

Figure 6-10. Underground cable protection provided by the sheath.

Conductors

The conductors are usually made of copper or aluminum and are stranded, except for the smaller sizes, to make the cable flexible. As the conductor has to support practically no weight, it is usually made of soft-drawn copper. The size is specified in circular mils or in American Wire Gage numbers exactly the same as for overhead conductors.

Cables can be either single-conductor or multiple-conductor (see Figure 6-11); that is, one, two, three, or four conductors may be enclosed in a single sheath. Putting two or more conductors into one sheath is an economic measure, as only one cable is handled rather than several where they are installed in a trench or duct, and only one sheath is necessary. Where the cable is to be tapped rather frequently, as in service connections, secondary or primary distribution mains, and street lighting facilities, single conductor cables are generally used because of the ease in making the necessary splices. Multiple-conductor cables are usually used for mainline portions of primary distribution circuits, or transmission lines, where few or no taps or take-offs exist.

Insulation

Plastics occupy a prominent place in the field of insulation for cables operating at low and moderate voltages (up to 69,000 volts), though rubber-mineral compounds may be used for special applications. For cables of moderate and high voltage application (from 69,000 volts upward) oil-impregnated paper tape is widely used. Plastic insulation is applied to the conductor under pressure. Paper insulation is wrapped around the conductor and impregnated with oil before the sheath is applied. Sheaths may be of lead, though more frequently, plastics are being used for this purpose.

Armor

Where cable is buried directly in the ground, it is often constructed with armor which may consist of galvanized-steel wire wrapped around the sheath. While this armor primarily serves to protect the cable from mechanical damage during and after installation, it can also serve as a return or ground conductor as part of the electric circuit. This is especially true in the case of lower voltage service or secondary cables.

Submarine Cable

In crossing bodies of water, the ordinary plastic or lead covered cable is usually protected with a wrapping of tarred jute, and armored with galvanized-steel wire (see Figure 6-12). The armor wire is usually more substantial than that for cables buried in the ground, and serves also as mechanical reinforcement during installation.

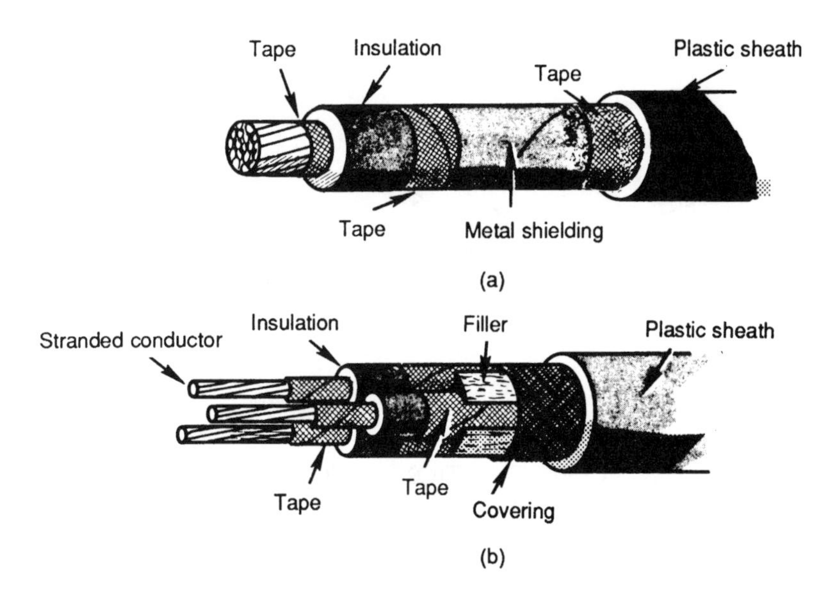

Figure 6-11. Types of cable. (a) Single-conductor cable and (b) three-conductor cable.

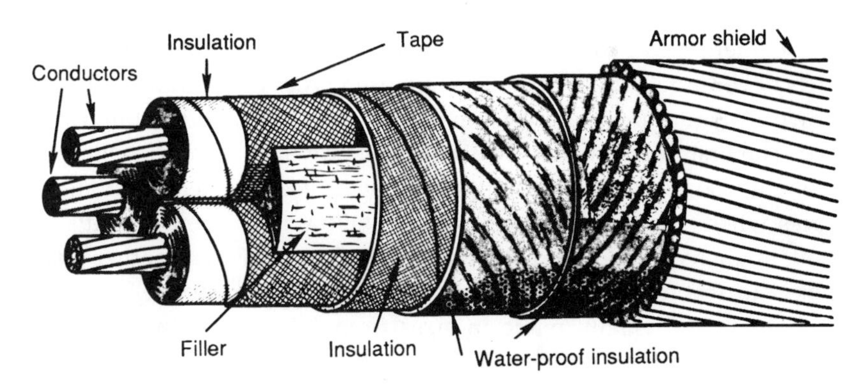

Figure 6-12. Submarine cable.

Where submarine cable is not likely to be disturbed, it is laid directly on the bed of the body of water, but where the current or navigation is likely to disturb it, the cable is laid in a protective trench blown or dug in the stream bed. Special anchors are sometimes used so that the force of a vigorously changing tide or moving water cannot shift the cable. Sometimes, these cables may be encased in concrete (see Figure 6-13 in the section that follows).

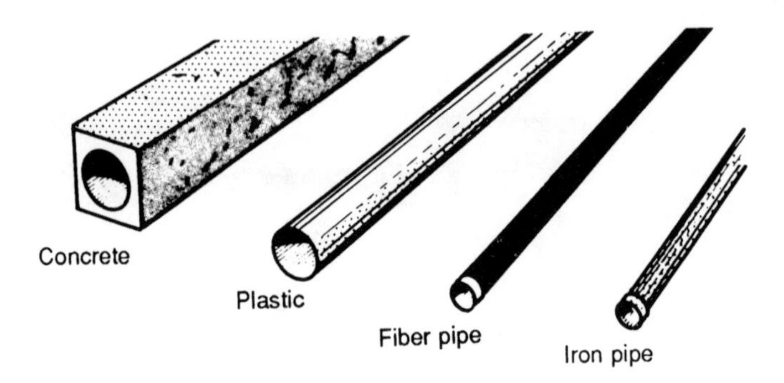

Figure 6-13. Ducts.

UNDERGROUND CONSTRUCTION—DUCTS

Cable in Conduit

Installing cables in underground conduit or duct systems is a more complex and costly operation than burying them directly in the ground.

Conduits

Underground cable is carried and protected by different types of ducts or conduits. The most commonly used types of conduit or duct are shown in Figure 6-13: precast concrete, plastic, fiber, and wrought iron pipe. The first three are usually used because of their lower cost. Because wrought iron pipe is comparatively costly, its use is generally limited to places where the space for conduits is shallow and where rigidity and strength are required. Conduits installed under roadways or other places subject to severe loading are sometimes encased in concrete.

Ducts are usually made in sizes from 1 inch to 6 inches (inside diameters), ranging in steps of 1/2 inch, that is 1, 1-1/2, 2, 2-1/2, and so on. The size of the duct depends on the size of the cable to be installed at the present or on any probable size to be installed in the future. It should always be large enough to make cable installation as easy as possible.

Very often, in installing conduits, spare ducts as shown in Figure 6-14 are provided for future use. In the illustration, the white-capped ducts are in use; the black ones are spares.

In installing the duct run (see Figure 6-15), it should be kept free of bends as much as possible so as to simplify the pulling in of the cable. For the same reason, the duct runs between manholes should not be too

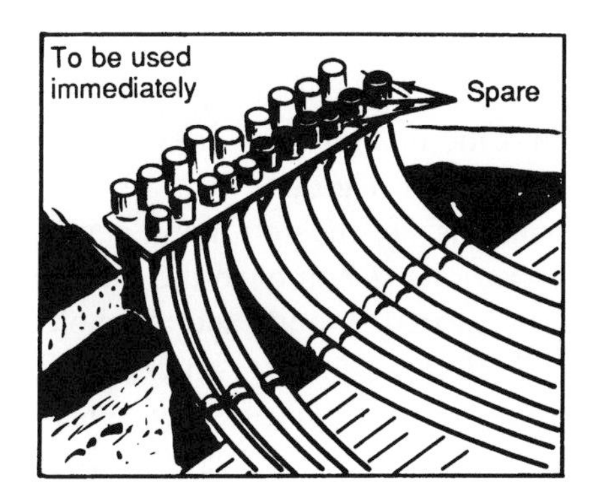

Figure 6-14. Preparing an installation of plastic conduits.

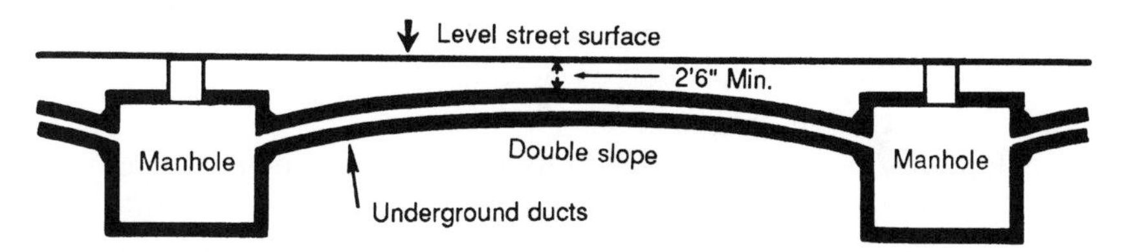

Figure 6-15. A method of grading a conduit between manholes.

long. If bends are necessary, they should be constructed with the greatest possible radius. The ducts should also be graded slightly, that is, they should slope gently toward one or both ends, to allow any water that may seep into them to drain off.

UNDERGROUND CONSTRUCTION—MAN HOLES

Manholes

A manhole, more accurately termed a splicing chamber or cable vault, is an opening in the underground system where workers may enter to install cables or other equipment, and to make connections and tests.

Manholes provide test points for the various circuits carried through them. They permit grouping the circuits as desired by the engineer, and they make emergency connections and maintenance possible and efficient.

Manholes are placed in underground systems to facilitate cable pulling. The distance between manholes depends on many conditions, but they are seldom over 500 feet apart. In built-up areas, they are usually located near street intersections. In addition to accommodating the cable or other equipment to be installed, they should be designed so that the cables may be readily racked around the manhole and they should allow sufficient space for a splicer to work (see Figure 6-16).

The type, shape, and size of a manhole depends on many conditions—location of manhole, whether at an intersection or in the middle of a block; size and type of cables to be carried through the manhole; and location of gas or water pipes, or other subsurface structures. It should be deep enough to provide working head-room (see Figures 6-17 and 6-18) and drainage for the duct lines. The dimensions of a manhole are also determined by the number of ducts entering the manhole, the different levels on which they enter, the type of cable to be racked, and the length of the splice, as well as by other types of equipment to be installed (switches, transformers, etc.).

Manholes are generally built of reinforced concrete or brick and the covers are made of steel. The section leading from the street to the manhole proper is usually called the "chimney" or "throat" (see Figure 6-19).

Both manholes and ducts are often precast into the sizes and shapes desired and installed as integral parts of the underground system. This is often the quickest, most economical way to install under-

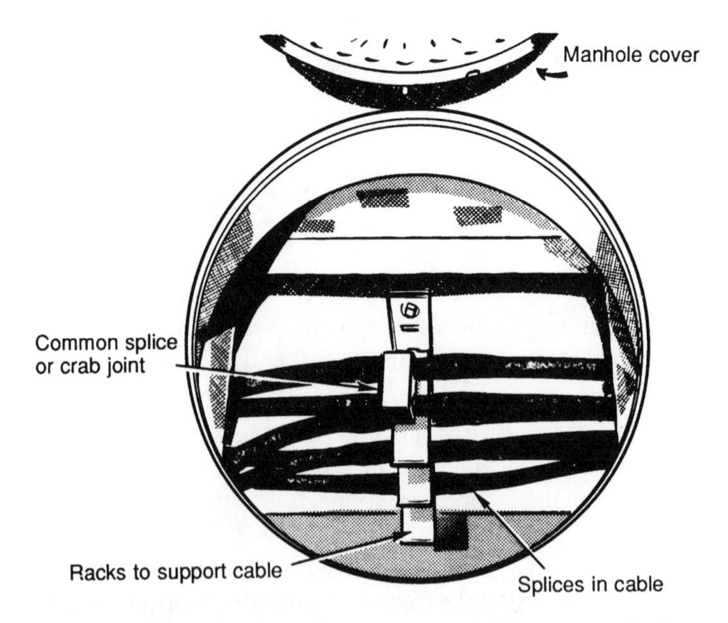

Figure 6-16. Cables spliced and racked in a manhole.

ground facilities, eliminating the inconveniences that often accompany the field pouring of concrete, especially in crowded streets and avenues.

Drainage and Waterproofing of Manholes

Water can accumulate at the bottom of manholes. If the soil is normal, natural drainage will easily occur. However, if there is less than

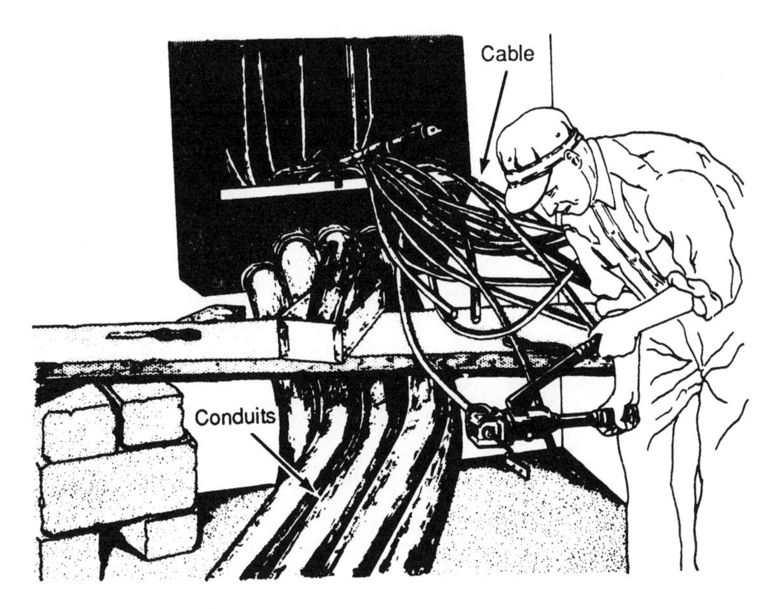

Figure 6-17. A manhole must provide ample headroom for workers.

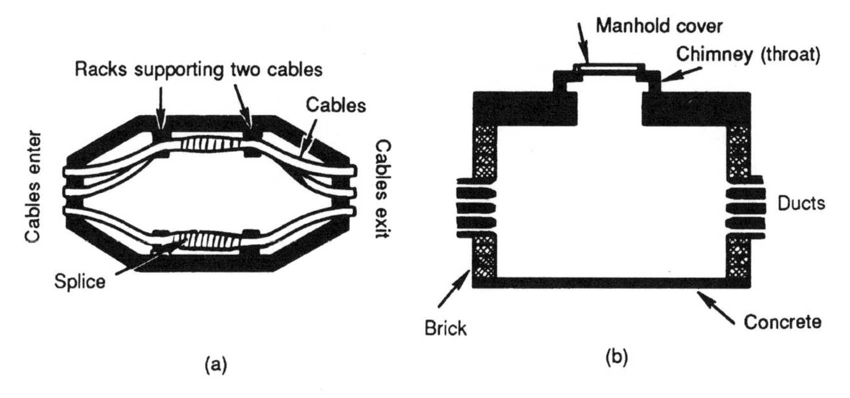

Figure 6-18. Cross-sectional views of a straight manhole. (a) Top view and (b) side view.

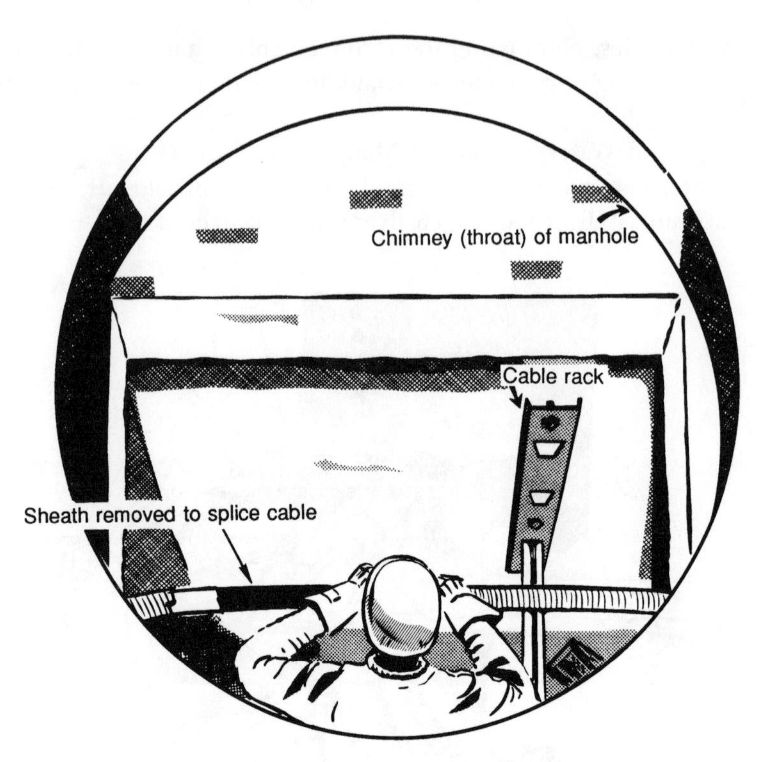

Figure 6-19. Looking down a manhole's "throat" or "chimney"; a worker splicing a cable.

perfect soil, some form of drainage should be provided. Where sewer connections can be made conveniently, this is usually done. In most cases where the bottom of the manhole is below the natural water table or where the earth will not support the manhole structure on the wall footings alone, it is good practice to provide a concrete floor and a sump (or well) where the water can collect and drain off into a storm sewer or some other part of the ground.

In many cases, the bottom of the manhole is excavated and filled with stones to make a dry well. (See Figure 6-20.)

There are areas where the soil is so moist and the water table so high that it is wise to make the manholes as waterproof as possible. Along the waterfront in cities, for example, the manholes are constructed of water-proofed concrete. For extra protection, the concrete is sometimes painted with a waterproof paint.

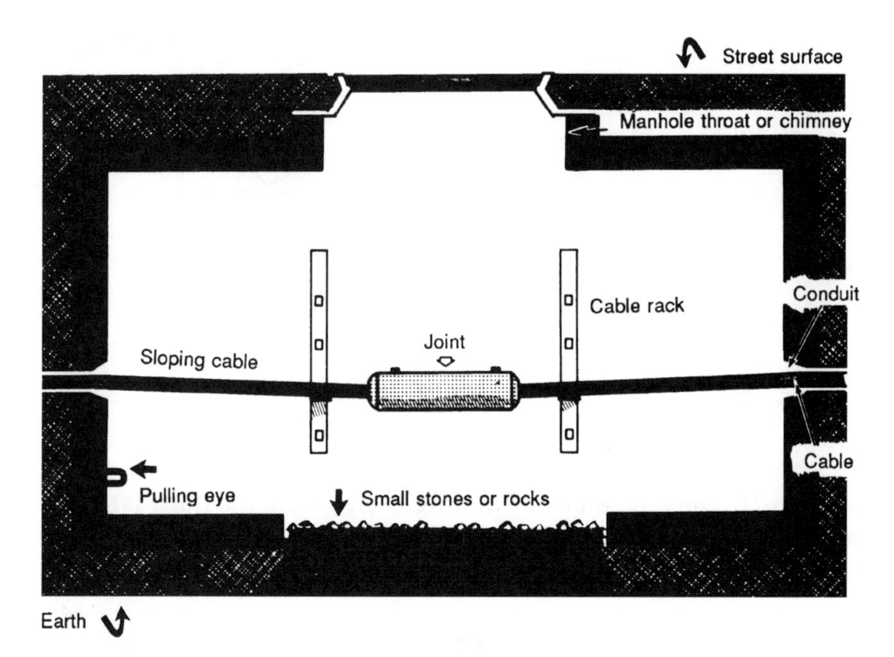

Figure 6-20. Typical manhole with dry well for drainage of rain water.

UNDERGROUND CONSTRUCTION—CABLE INSTALLATION

Pulling Cable through a Conduit

When installing a cable into an underground conduit, it is possible that the engineer can choose from several conduits. If so, the choice is made with the following stipulations in mind: (1) avoid unnecessary cable crossings in the manholes, (2) avoid obstructing other empty ducts, and (3) keep the cables cool and safely distant from other cables.

After the conduit has been chosen, it must be cleaned. At this point the pulling line is often installed in the conduit. However, often the pulling line is placed in the conduit at the time of its original installation. Manila or steel wire rope are commonly used materials for pulling lines.

Pulling the cable through the conduit requires a reel of cable properly placed at the feeding end and a winch, capstan, or truck at the pulling end (see Figure 6-21). Different companies have different means of attaching the conductor or cable to the pulling line.

To protect the cable from injury during installation, the cable is usually greased lightly (about 100 lb of grease per mile of cable). The

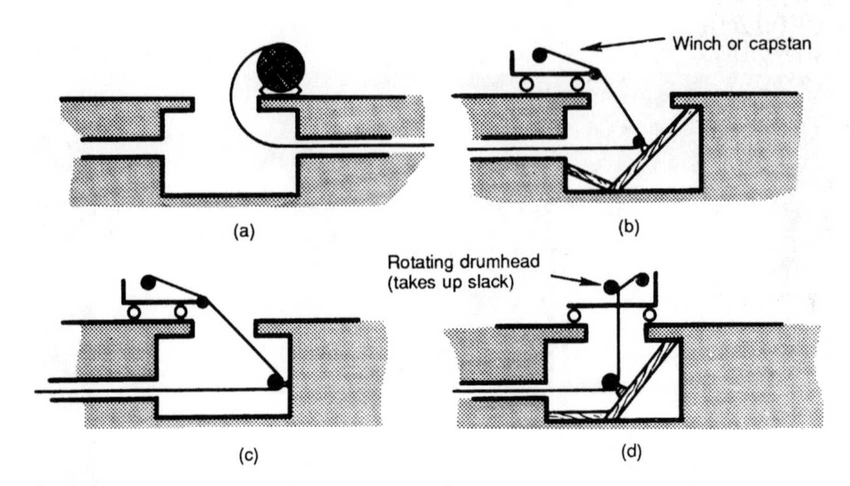

Figure 6-21. Some methods of pulling cable through a duct. (a) Feeding cable into a duct, (b) simple rigging-blocks and wedges, (c) eye imbedded in manhole well, and (d) use of drumhead.

feeding end of the conduit is smooth and it is good practice to pull the cable slowly.

Joints or Splices

Sections of cables installed in underground ducts are connected in the manhole to form a continuous length. This connection is called a *joint* or *splice*. Whenever two cables are spliced together, the connection is called a *straight splice*. It is possible to connect more than two cables together; these are known as three-way, four-way, and so on, splices.

Where joints or splices are made in cables buried directly in the ground, the splice may be also buried directly, or may be protected by a splice box or chamber surrounding the joint.

Where cables having plastic insulation and sheathing are used, the splices may utilize prefabricated insulation covers, and the entire operation simplified, as compared to cables having other insulation and lead sheaths.

Connections-Terminals

In some instances, particularly where plastic insulated cables of underground residential systems are involved, cables may be connected together through a completely insulated terminal block containing a number of stud connectors, the entire assembly covered with molded insulation, as shown in Figure 6-22(a). The conductor is connected to the stud and the insulation and sheath of the cable are taped to the insulation and molded protection of the insulated terminal block. The terminal may also consist of molded load-break elbows and bushings, as depicted in Figure 6-22(b). The conductor is connected to an insulated stud that fits into an insulated receptacle. The conductor may be disconnected safely by means of an insulated hook stick that pulls the stud from the receptacle. The molded terminal may be combined with similar type terminals of transformers and other equipment. These disconnecting devices may serve as test points in finding and isolating faults, in restoring service after an interruption, and lend flexibility rearranging circuits as well as energizing and deenergizing primary circuits and equipment.

Cable Racking in Manholes

As the cable expands and contracts, the sheath may crack. That's why it is very important to rack the cables in the manhole so that this movement can take place safely. The best way to accomplish this is to provide a large reverse curve in the cable passing through the manhole. This reverse curve, consisting of one or two large radius 90-degree bends, enables the cable to take up the expansion movements and minimizes the danger of the lead sheath cracking or buckling. This arrangement also minimizes the relative movement between the cable and the

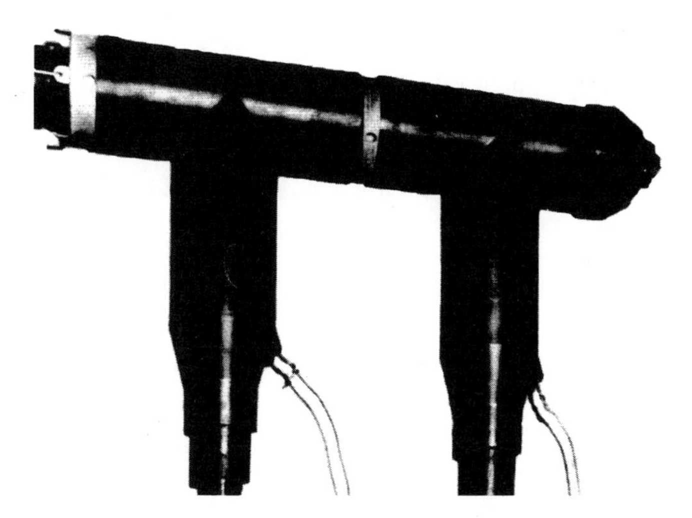

Figure 6-22. (a) Premolded typical "T" splice assembly. (Photo courtesy of Elastimold, a unit of Thomas & Betts Corporation.)

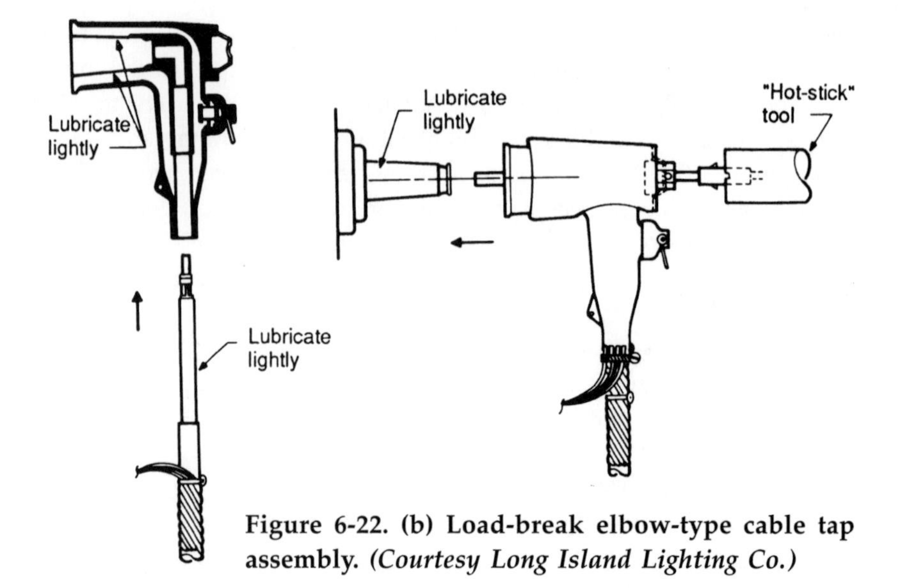

Figure 6-22. (b) Load-break elbow-type cable tap assembly. ***(Courtesy Long Island Lighting Co.)***

joint at the point where the joint is wiped to the cable sheath.

Splices in manholes are usually supported on racks mounted on the walls. As the splice is a vulnerable spot, the cable ends must not hang on it; this would tend to pull the splice apart.

Cables should bear some identifying mark (see Figure 6-23) in the manholes through which they pass. This is usually in the form of a metal tag tied to the splice. The tags indicate the size, voltage, feeder designation, and other pertinent information.

Arc-Proofing Cables

Where more than one high-voltage cable passes through a manhole, the splices and cables are fire-proofed with a covering of sand and cement, or some fire proof material. This is to prevent spread of fire or explosion, should a splice or cable fail, to other cables, with consequent damage.

First, the cables to be fire-proofed (or arc-proofed) (Figure 6-24) are formed and racked into their final position. In the situation described, three single-conductor cables are being arc-proofed together. These are tied together with twine. If splices are present, the twine is wound only at the ends, not over the splices. Now two layers of special-reinforced tape are applied around the bound-together cables. The second layer must be carefully wound so that the joints occur just between the joints of the first

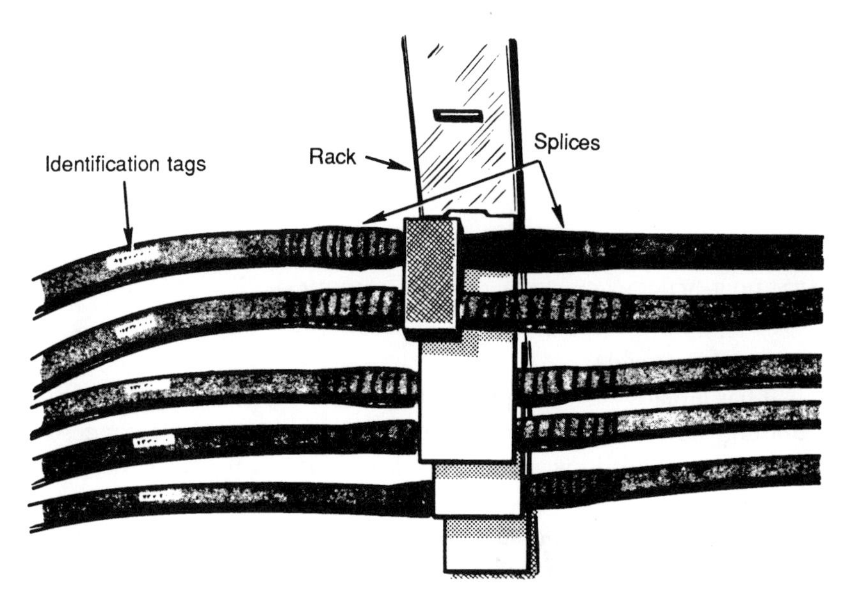

Figure 6-23. Tagged cables supported on racks in a manhole.

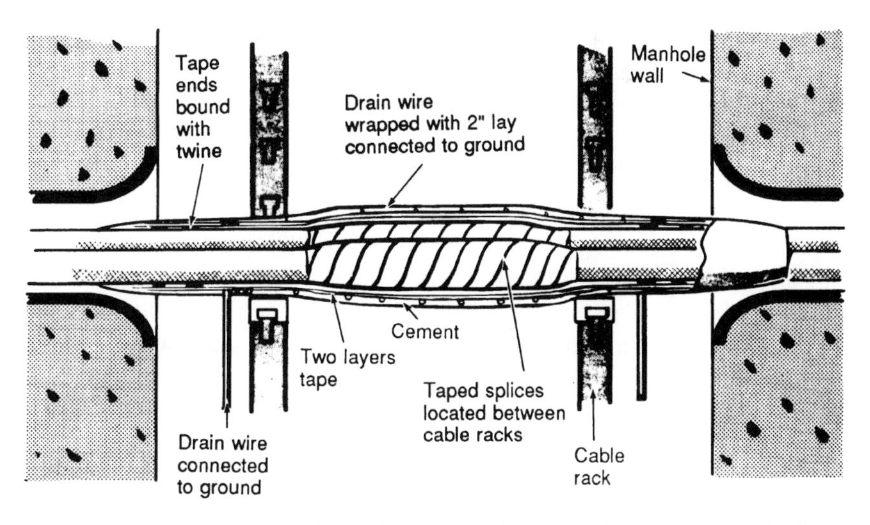

Figure 6-24. Arc-proofing single-conductor cables in a manhole.

layer. Both ends of each layer must be secured with twine.

For safety, a drain wire is wound around the tape and connected to ground. This is usually a solid soft-drawn #6 copper wire mesh, wound in the opposite way to that of the insulating tape, with 2 inches between turns. At each end, the turns are wrapped closely and enough left over for connecting to the ground. The insulating-cement (about 3/8-inch thick) is rubbed into the insulating tape, which is wet and smoothed off the rubber gloves or a trowel. The last step is to connect the drain wire to ground.

UNDERGROUND CONSTRUCTION—RISERS

Risers and Potheads

It has already been mentioned that the utility company only resorts to underground construction where it is necessary, where congestion, maintenance conditions, or appearance make overhead construction impractical.

Therefore, in a long-distance transmission or distribution system, it is possible that once the conductors have passed underground through some crowded area, they may be able to connect to overhead lines in a more scattered area.

To make this underground-to-overhead connection, the cable is led out of the manhole and up the side of the pole through a curved length of pipe. The cable terminates in a pothead or weatherhead (see Figure 6-25). The whole assembly is known as a riser.

A weatherhead is a simpler underground-overhead connector; but it can only be used for low voltages, that is, to connect underground secondary cables to overhead secondary or service wires.

A pothead (shown in Figure 6-26) contains the connection between an underground cable and an overhead wire. Inside the pothead, all the conductors wrapped into one cable are separated into two, three, or more conductors for overhead attachment through a porcelain-encased terminal for each conductor. These terminals or arms have petticoats the same as insulators do and serve the same purpose. The sheathed cable is attached to the pothead by a wiped joint or by a clamping device. The overhead wires are connected to the terminals. Before this connection is made, the pot is filled with a liquid insulating compound which must be allowed to cool before the pothead can be moved.

On distribution lines where the line equipment must be handled while the line is alive, a disconnecting-type pothead (see Figure 6-27) is used either for single or multiple-conductor cables (see Figure 6-28). These potheads also consist of a pot (or container of insulating com-

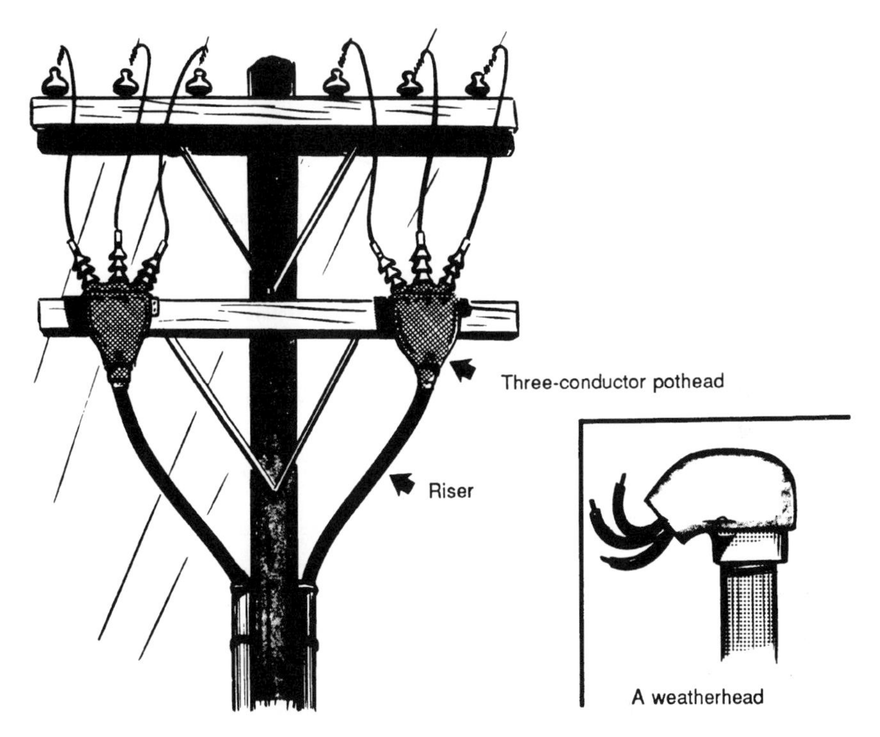

Figure 6-25. Two potheads connecting two underground cable lines with two overhead lines.

pound) and of porcelain tubes for the outlets to the overhead lines. They differ from ordinary potheads in that the tubes are covered with a weatherproof cap provided with a terminal which is constructed in such a manner that when the cap is removed, the circuit is interrupted. The cable terminals thus permit the opening of the line. Such an arrangement is useful in locating trouble. Potheads also provide a means of restoring service to a partly faulted line by transferring the supply to an adjoining main.

With the creation of plastic insulated and sheathed cables, connections may be made directly to the overhead conductors. The insulation about the underground conductor is tapered at the end of which a hot-line clamp may be mounted. Conical rain shields are sometimes installed at the point where the conductor is attached to the clamp. The clamp may be operated by means of a hot-line stick in connecting and disconnecting the cable conductor and the energized overhead conduc-

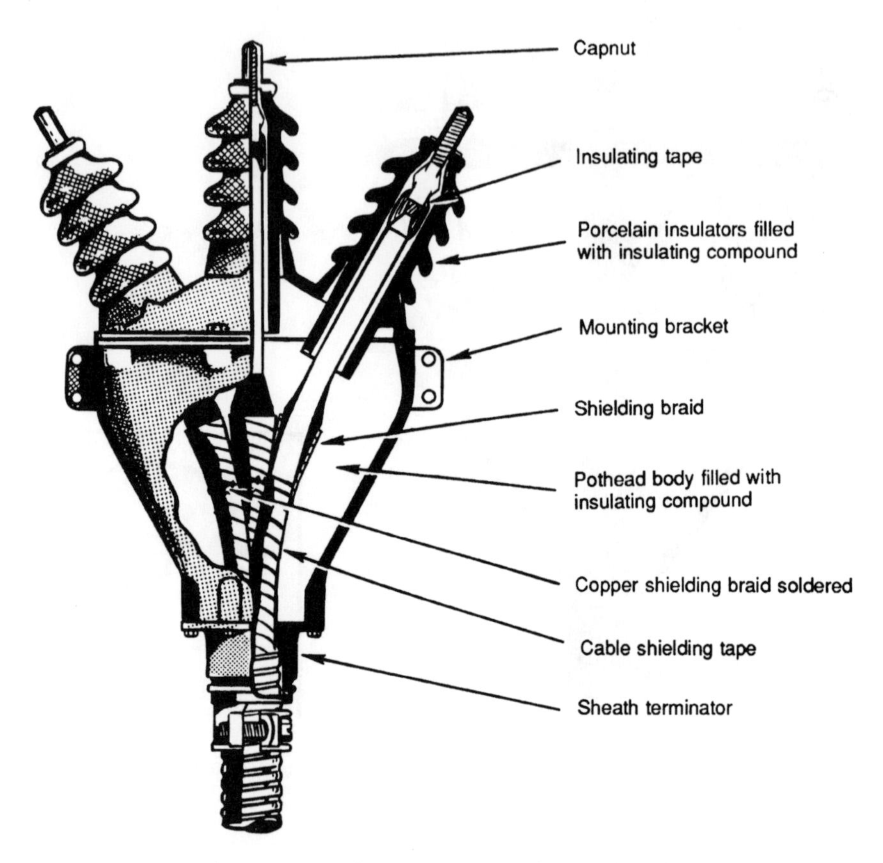

Figure 6-26. Cross section of a pothead.

tor. Each conductor of a multiconductor cable is handled individually. The economy, ease, and flexibility of operation compared to pothead installations are obvious.

OVERHEAD CABLE

Overhead Sheathed Cables

For appearance or to avoid congestion, it is sometimes necessary to continue circuits by means of sheathed cable suspended on poles. This cable is supported by means of rings attached to a tightly strung, heavy steel cable (see Figure 6-29) called a *messenger.* More recently, instead of rings, the cable is supported by a wire wrapped around both the cable

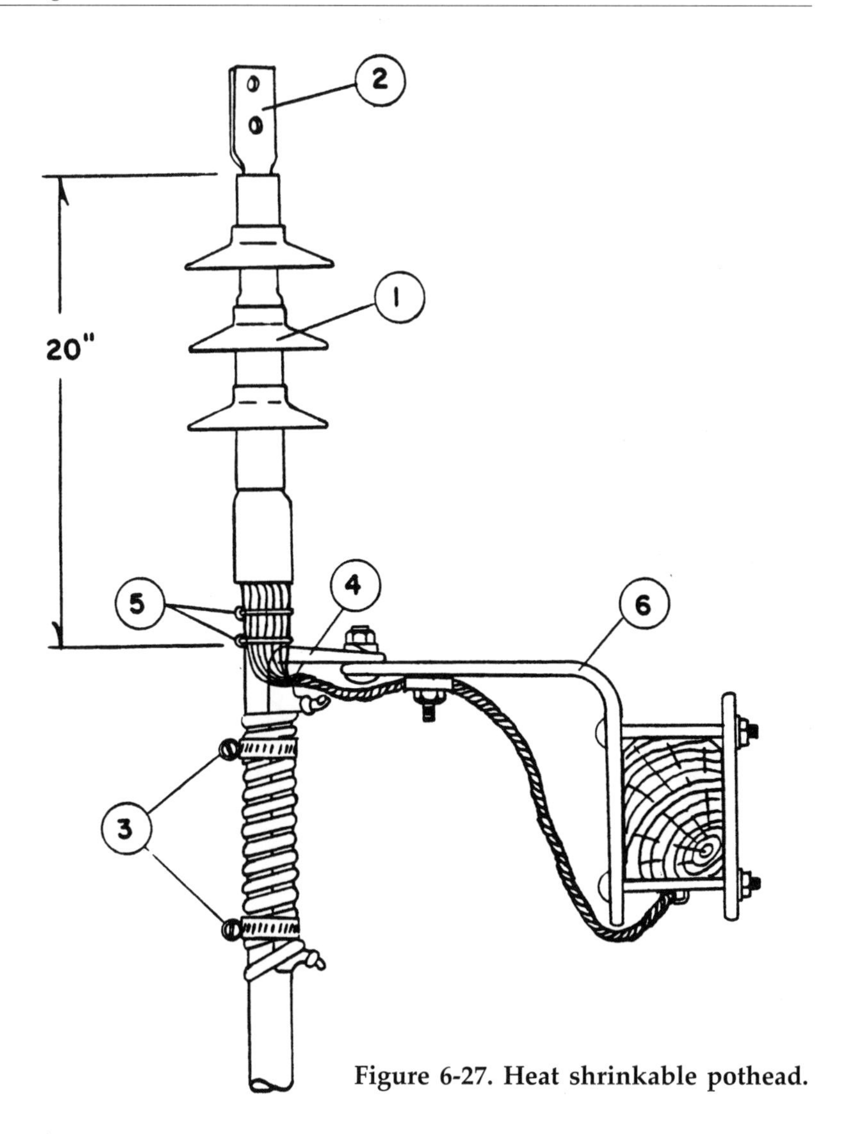

Figure 6-27. Heat shrinkable pothead.

and the messenger by a spinning machine (see Figure 6-30). Bad tree conditions are another reason for overhead sheathed cables.

Because of the pull imposed on the poles by this tightly strung messenger, and because of the weight of the cable, the pole line must be very strongly constructed and heavily guyed. Since there is enough insulation between the conductors and the lead sheath, no insulators are used for cable work. This cable expands and contracts just as do the conductors in an overhead system and it will sometimes rub against the

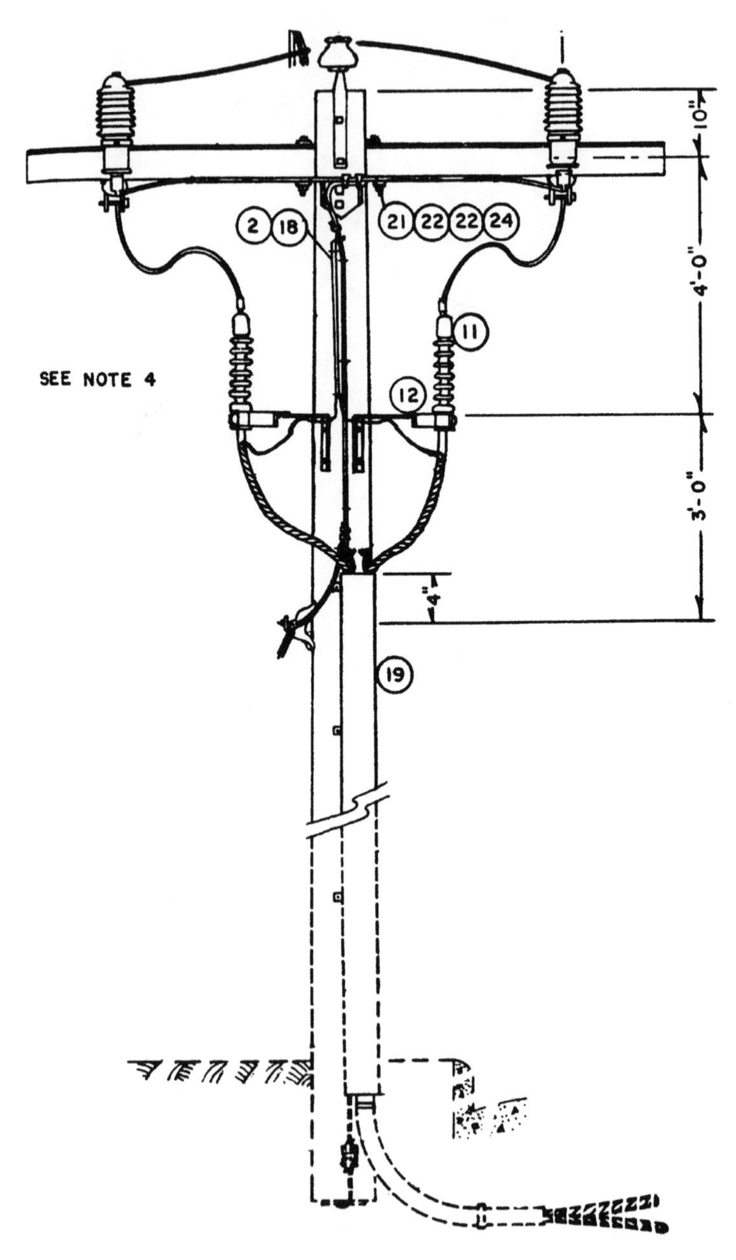

Figure 6-28. Two single-conductor potheads.

pole. To prevent damage from this rubbing, the cable is sometimes "bellied" out around the pole, as shown in Figure 6-31(a) and (b). Otherwise, some type of mechanical protectors such as steel plates are used.

REVIEW QUESTIONS

1. What is the advantage of burying power cables and parkway lighting cable?
2. What are the principal materials used to insulate underground cables?
3. What are the three main parts of a sheathed cable?

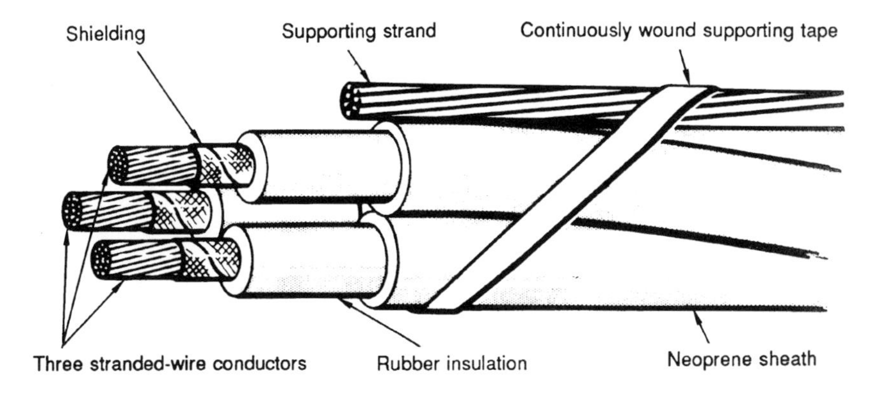

Figure 6-29. A self-supporting aerial cable.

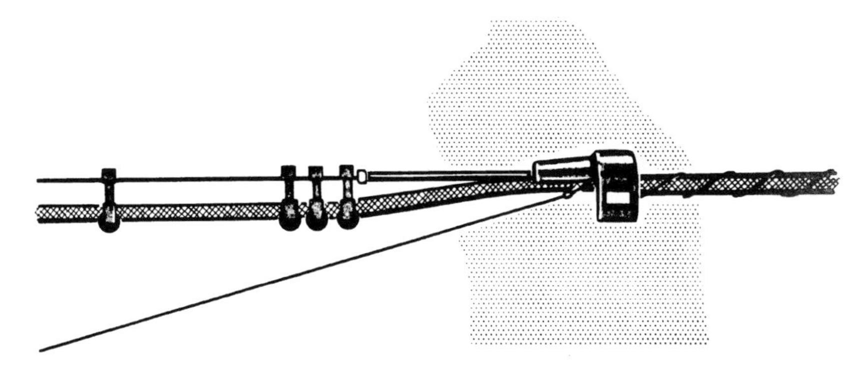

Figure 6-30. A spinning machine wraps wire around a cable and its messenger.

4. Name the various kinds of conduits commonly used.
5. What are the functions of a manhole? What are some of the advantages of precast ducts and manholes?
6. Why are underground cables sheathed?
7. What is the purpose of arc-proofing cables and splices in a manhole?
8. What is a riser? A pothead? A weatherhead?
9. What additional materials are used for submarine cable?
10. How are overhead cables suspended between points of support?

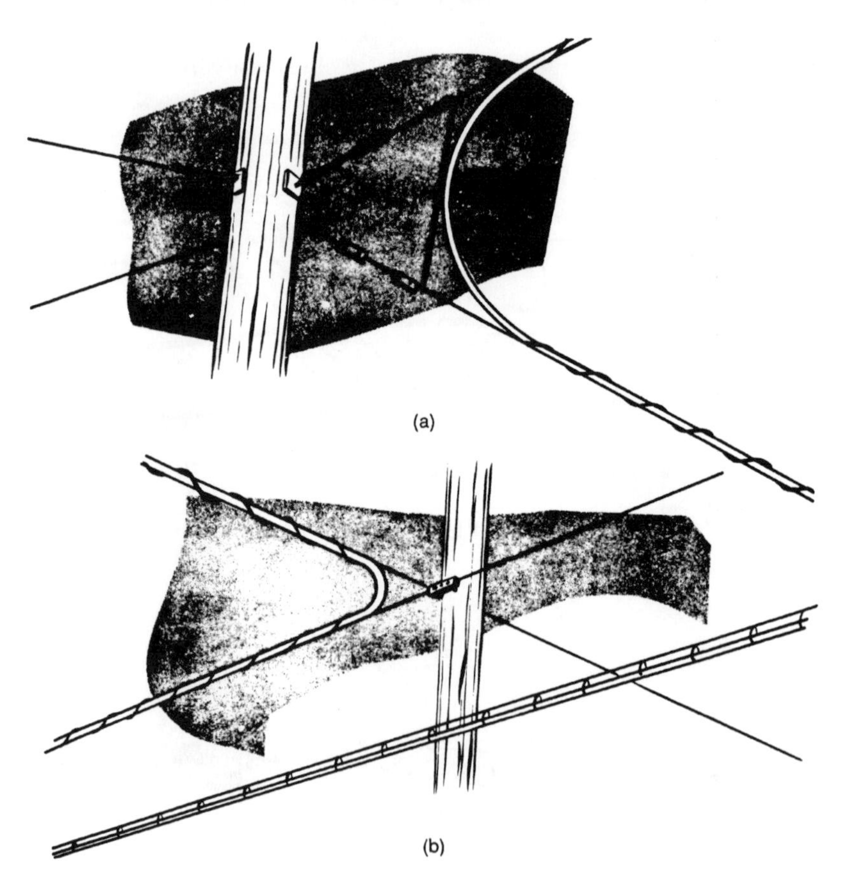

Figure 6-31. Cable "bellied" out around a pole.

Chapter 7

Service Factors

SERVICE FACTORS

If all the lights, irons, washing machines, refrigerators, television sets, motors, and other equipment in all the consumers' premises were to be turned on at one time, it would be impossible for the utility company to supply the energy necessary to run all these appliances. For a company to install generators, transformers, transmission and distribution lines, and other equipment to provide for this total connected load would be extremely uneconomical since all this equipment will not be operating at the same time. Therefore, studies are made of the personal habits of people and the routines of households, offices, stores, and factories, to determine what probable load the utility company will have to supply at any one time (see Figure 7-1). Although holidays, the season of the year, the weather, and many other factors contribute to make such studies difficult, this effort is necessary to keep the cost of electric service within everyone's range. Figure 7-2 illustrates typical peak and off-peak load periods.

Maximum Demand

The amount of electricity a consumer uses at any given moment is called his "demand." It varies from hour to hour. The highest value which the demand reaches is called the "maximum demand" (see Figure 7-3). It is evident then that this maximum demand for electricity (and not the actual consumption of electricity) determines the size and type of equipment necessary to supply the consumer. The higher the maximum demand, the larger and more expensive must be the apparatus necessary to carry it, from the watt-hour meter all the way back to the power house.

The maximum demand may exist for only a very short while or it may last for a considerable period. For practical purposes it is taken as the highest 15-minute (or half-hourly or hourly) demand during the billing period, usually 1 month.

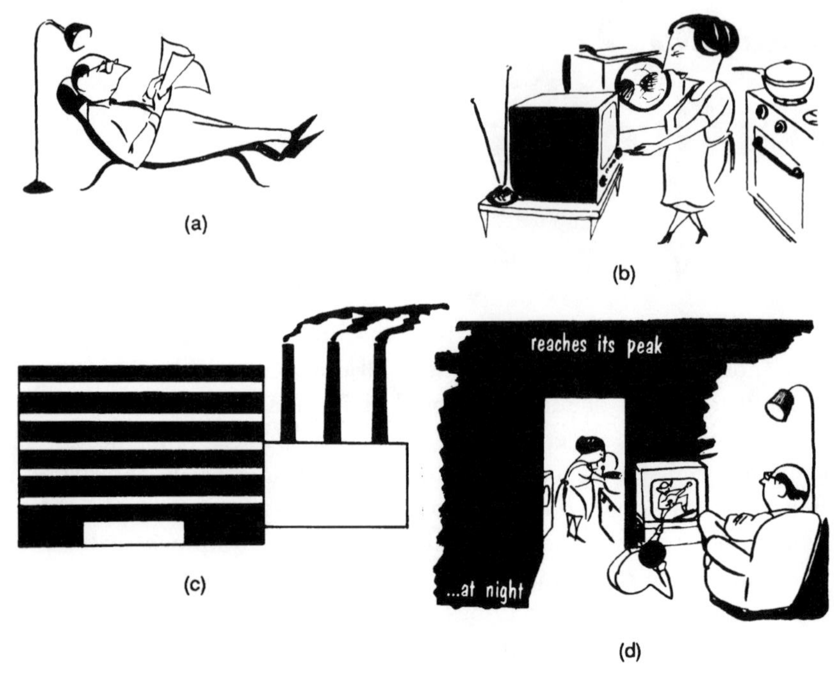

Figure 7-1. Service factors. Very low demand: (a) Consumer A has only one lamp on—no other appliances running. Higher demand: (b) Consumer B is cooking, washing, and running a TV. (c) The factory reaches its maximum demand during the day. But most factories are closed at night when (d) the domestic demand reaches its peak.

Load Factor

On this basis, it is clear that the consumer with a large maximum demand requires a larger investment on the part of the utility company than one with a small maximum demand. But, before any conclusions can be drawn as to their relative value as consumers, something must be known about their consumption of electricity as well as their maximum demands. If a consumer approaches this maximum demand often during the day, his or her consumption of electricity will be large. If the average demand in a given period of time is taken and compared to the maximum demand, the percentage is an estimate of the extent to which the facilities are used. This comparison, known as the load factor (see Figure 7-4) is actually the ratio between the average demand and the maximum demand in percent form.

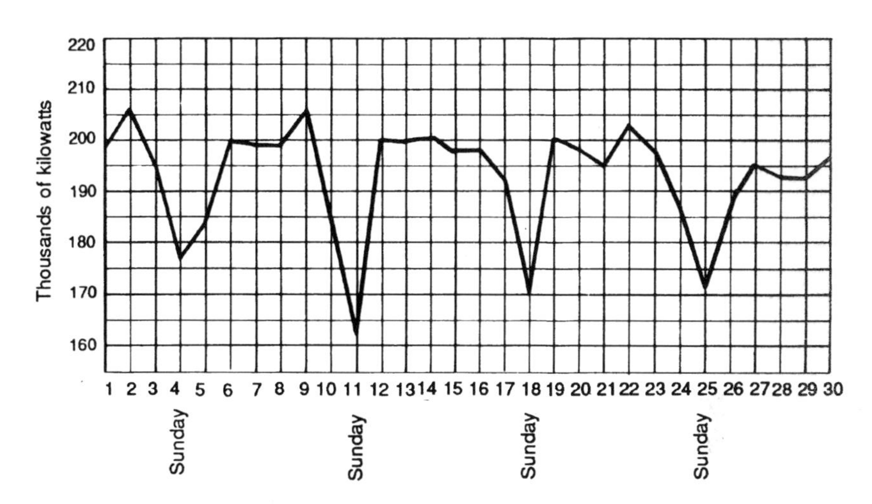

Figure 7-2. Daily peak load for a typical month. How many kilowatts a typical utility company supplies consumers in an average month. Note consistently low peaks on Sunday.

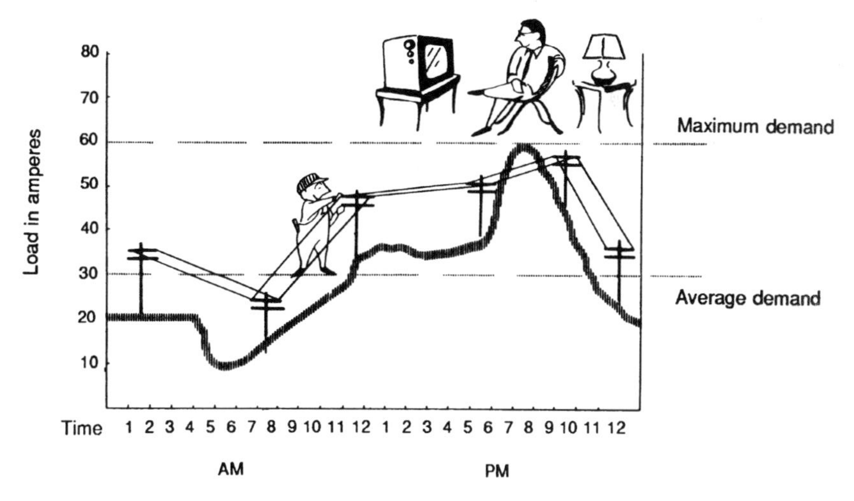

Figure 7-3. The utility company must be equipped to meet the consumer's maximum demand.

Even though the capacity of the installation is determined by the maximum demand, the load factor is of great importance because the use (and consequently the revenue) from the installation is governed by the average load. Considering the daily load factor, it is easy to imagine

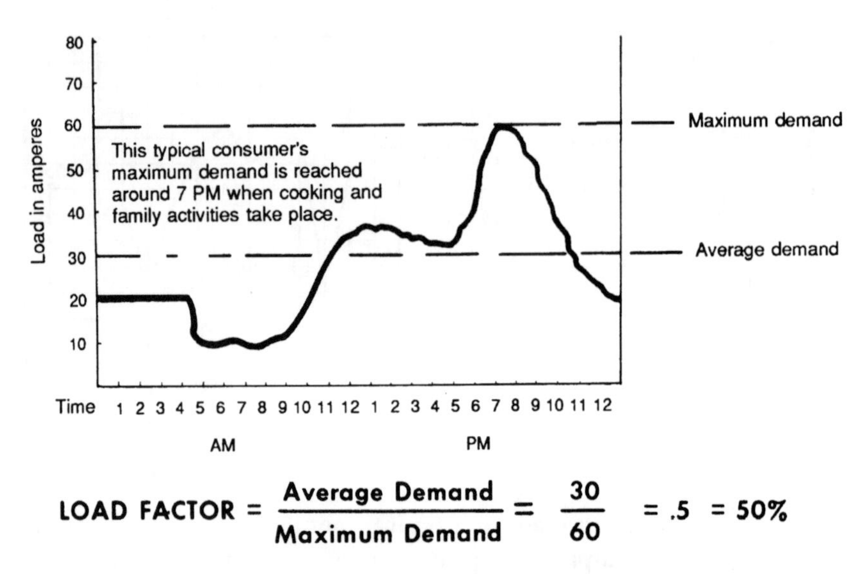

Figure 7-4. A consumer's load demand varies.

two consumers using the same maximum demand but having entirely different average loads. For instance, one may use this maximum demand steadily for 24 hours—a 100 percent load factor. The other consumer may use that demand for only an hour while the rest of the day only 10 percent of the maximum demand is used. Both consumers require the same investment in capacity while using that capacity in an entirely different manner.

The two consumers shown in the comparative graphs in Figure 7-5 have the same maximum demand. However, one uses his or her maximum for 1-1/2 hours—the other for 15 minutes. Their average demands vary by 15 amperes. Their load factors vary even more considerably, consumer A having 60 percent and consumer B only 30 percent.

Similarly, consider two houses; one used as a year-'round residence, the other for only the three summer months. Assume, too, that both have the same average demands and the same maximum demands; their load factors will also be the same. However, it is evident that one will use its installed facilities four times as fruitfully as the other.

Another factor often used is the *use factor.* This is the same as the load factor but the capacity installed is used instead of the maximum demand. It is usually less than the load factor.

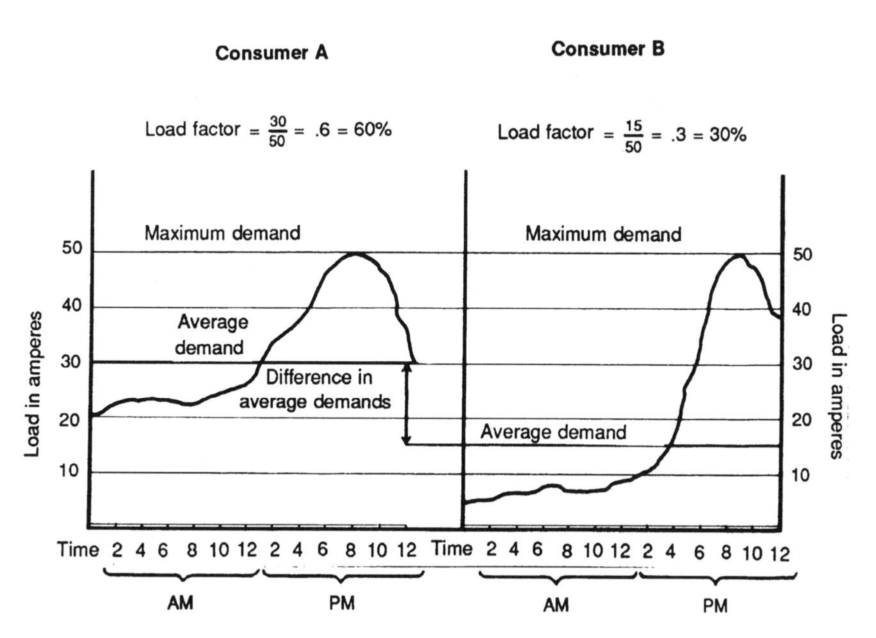

Figure 7-5. The load factor indicates how two consumers with the same maximum demand use their facilities in different quantities.

Load Characteristics

Each use of electrical energy has its own characteristics. The street lighting load—while it is on—has almost a 100 percent load factor. The domestic load (for homes) has a very high evening peak and varies enormously depending on the appliances used (Figure 7-6). Industrial load curves vary with the specific industry but generally are typified by two peaks, one in the morning and one in the afternoon.

Annual load curves show less variation than daily load curves. The peak load for the year, referred to as the system peak, often occurs during the two weeks before the end of the year, around the Christmas season. With the widespread use of air-conditioning, this system peak often occurs during the summer months.

If the average daily load curves taken for a whole year are placed in a row from January to December, the result is the mountainous graph shown in Figure 7-7. This "mountain range" is a good indication of America's daily electricity-usage behavior patterns. Peaks are always higher for weekdays than for Saturday and Sunday.

There is a consistent slope from 12 *midnight* until 3 *AM* indicating

Figure 7-6. The graph indicates sleeping, working, eating, and playing habits of a consumer.

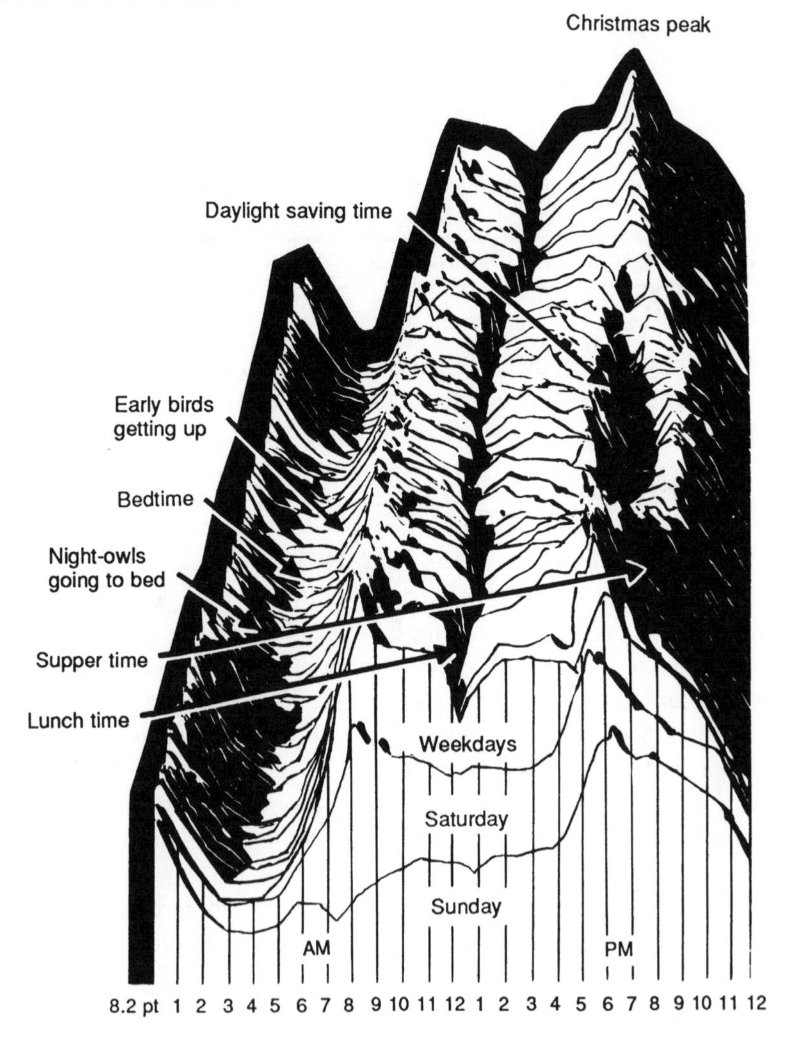

Figure 7-7. America's routine behavior. System demand graphs tell the story.

late workers and night owls going home. The curve is kept up during the night by the streetlighting load. Early risers account for the mountain rising again around 5 *AM*. Around 7 *AM*, the rise is sharp because of breakfast and transportation to work. By 9 *AM*, the peak of morning activity is reached. This slackens as transportation to work slows up.

Noon, being the usual lunch-hour, consistently causes a gulch in the graphs. Note that the gulch moves back an hour during daylight saving time. Daily peaks are almost always reached around 7 in the evening when family activity takes place.

This peak slips down around 9 *PM* as the family goes to bed. Daylight saving time causes an exception to this rule because people stay outdoors later in the day.

Weather and special events cause variations in this pattern. A summer storm will put the evening load at its winter peak. The evening load may be extended by some special radio or television program after 9 *PM*.

Diversity Factor

The time of the maximum demand of each consumer is important. Consider two houses with the same maximum demand but occurring at different times. If these two houses were fed from the same feeder, the demand on the feeder would be less than the sum of the two demands. This condition is known as diversity. If we compare the actual maximum demand on the feeder with the sum of the maximum demands of all the consumers supplied by the feeder (regardless of when they occur), the relationship, expressed in percent, is known as the *diversity* factor. The diversity factor can, therefore, never be less than one and may be as high as ten or more. The diversity factor of a feeder would be the sum of the maximum demands of the individual consumers divided by the maximum demand of the feeder (see Figure 7-8). In a similar manner, it is possible to compute the diversity factor on a substation, a transmission line, or a whole utility system.

Diversity factor and load factor are closely related. The smaller the load factors, the less chance of the maximum demands of two loads being simultaneous. The domestic load has the highest diversity factor; but the diversity factor decreases with the increase of appliances. Industrial loads have low diversity factors, usually in the neighborhood of 1.4, street light practically unity, and other loads vary between these limits.

The actual facilities installed by the utility company provide for the

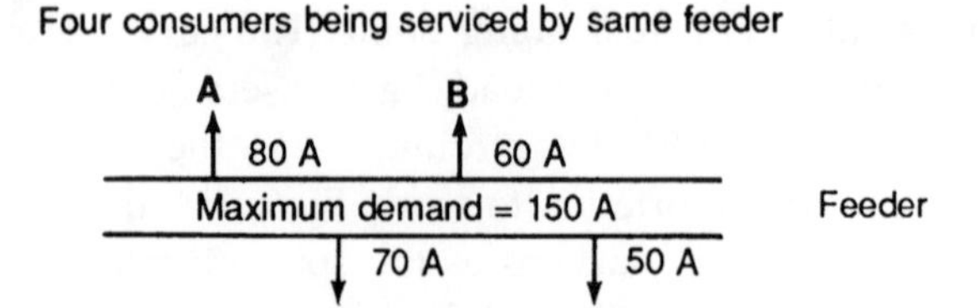

Monday: A Reaches maximum demand of 80 A
Tuesday: B Reaches maximum demand of 60 A
Wednesday: C Reaches maximum demand of 70 A
Thursday: D Reaches maximum demand of 50 A

$$\text{Diversity factor} = \frac{\text{Sum of total demands}}{\text{Maximum demand on feeder}} = \frac{260}{150} = 1.73 \times 100\% = 173\%$$

Figure 7-8. Diversity factor.

maximum *probable* loads, rather than maximum possible loads in any portion of the system. Thus, it will be seen that considerable savings in investment are brought about by taking advantage of all these factors.

Demand Suppression

It has been shown how the maximum demands not only of individual consumers, but also of their effect collectively on the maximum demand, impose on the conductors and equipment of a distribution system. Measures to reduce these maximum demands result in a lesser need for costly generating facilities with further reductions realized from decreased losses in both transmission and distribution systems. And many of the measures also include reductions in consumers' consumptions by replacement of lighting, heating, air-conditioning, and some power devices with more efficient units.

In general, the reduction in demands, sometimes also called *peak shaving, is* accomplished by dropping interruptible loads, and by controlling major load units by means of relays and circuitry that restricts their coincidental energization. Concentration on large industrial and commercial consumers not only produces larger results but may be more quickly and easily achieved. Similar measures are taken in the case of residential consumers. Major appliances such as clothes and dish washers, air conditioners, water heaters, and so on, may also be wired through relays so that they do not turn on at the same time. In some instances, they may be metered separately. These meters, as well as some of the relays, may be operated by radio signals or other type control at regularly scheduled time periods or predetermined demand levels. In all cases, promotional rates (for decreased consumption and demand) may be employed.

SERVICE WIRES

Services-Overhead

The electrical circuit between the company's mains and the consumer's wiring is called the *service.* A consumer's service or service connection is the set of wires that is tapped onto the secondary mains and is connected to the consumer's wiring. These wires are also known as the *service drop.* It is the last link of the path over which the electrical energy is brought to the consumer.

An overhead service (see Figure 7-9) consists of wires or cables extending from a pole carrying the main to a point on the consumer's building. The service may consist of two or more conductors. The con-

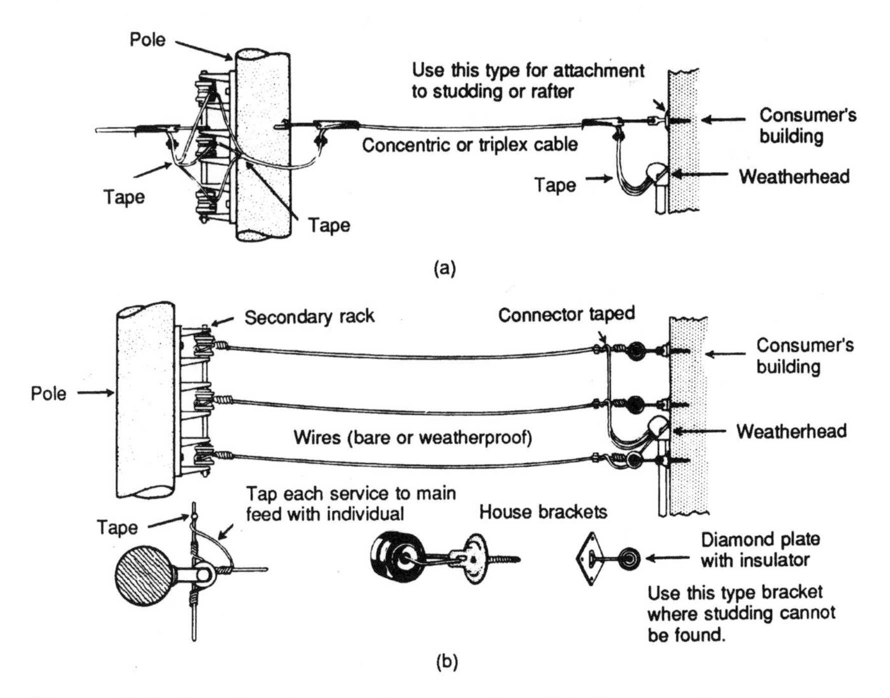

Figure 7-9. Overhead secondary service. (a) Concentric or twisted cable, and (b) open wire.

ductors may be covered with some insulation and contained together in a cable or twisted together. The conductors may also be bare or weatherproof wire. This latter type is known as an *open wire service* and in such installations the wires are supported by the insulators and brackets attached to both the pole and building. In any case, the conductors are usually supported so there will be no pull on the splice or connection.

Services—Underground

Underground services (see Figure 7-10) consist of plastic or lead covered cables which extend from the consumer's service point on his or her premises to the mains to which they are connected. These mains can be cables directly buried in the ground, or located in a manhole or splice box; or they may also be connected to overhead mains on a pole through a "riser" which runs down the side of the pole. Like overhead services, they may consist of two or more conductors which may be in the form of two or more single conductor cables or one multiconductor cable. These service cables are often installed in a steel or concrete duct, for at least that portion on the consumer's premises, for safety reasons.

Sizes of Service Wires

The size of the service conductors depends on the maximum demand of the particular consumer. In order to allow some margin of safety and to keep the number of different sizes of conductors to a minimum, certain standard sizes of wires are set up to supply certain ranges of demand (see Figure 7-11). These are given in Table 7-1. Remember that these sizes are selected not only to carry the load, but to do so without appreciable loss in pressure or voltage.

WATT-HOUR METERS

The electric meter is essentially a small electric motor, capable of considerable accuracy. It is connected in the circuit so that the number of turns of the disc is in direct proportion to the amount of electricity flowing through the circuit. The number of revolutions of the motor are counted and recorded on the dials through a system of gears. This system of gears and the dials constitute what is known as the *register.*

When a meter is equipped with just this register, it is known as a *watt-hour meter.* This is the type of meter usually installed in residences

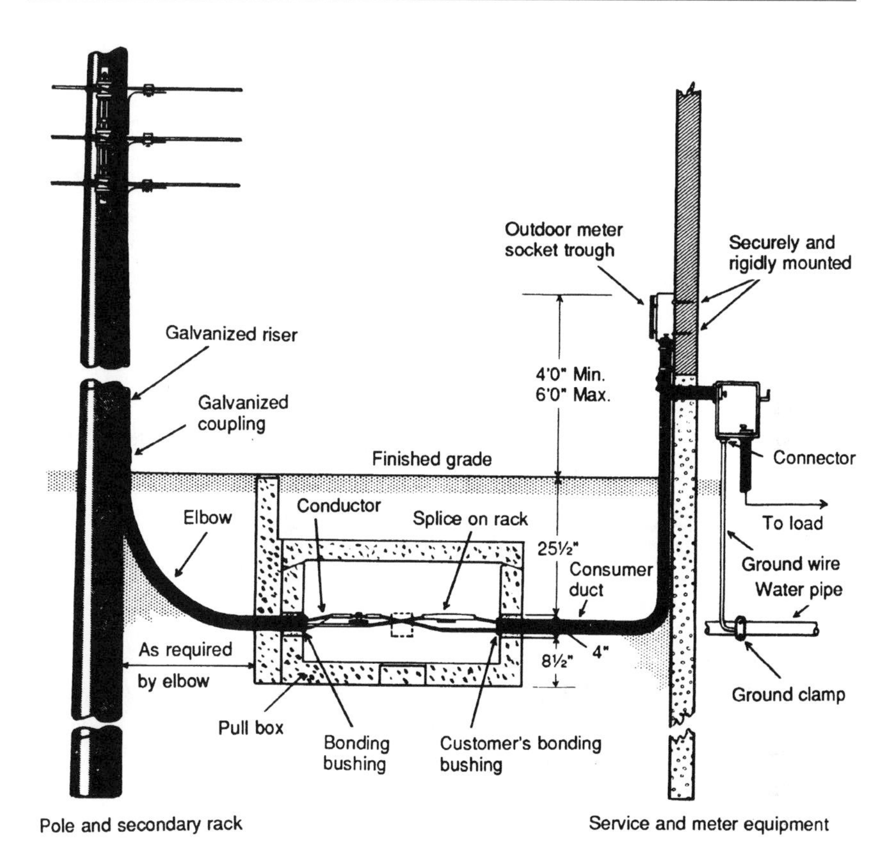

Figure 7-10. Layout for a typical underground service.

Table 7-1. Recommended Wire Size for Electric Services

Copper (AWG Size)	*Aluminum (AWG/Cir. Mils)*	*Load carry capacity in Amperes*	
		Overhead	*Underground*
8	6	0 to 75	0 to 40
6	4	0 to 100	0 to 50
3	1	100 to 200	Not used
2	0	100 to 200	50 to 90
0	000	200 to 300	90 to 120
00	0000	300 to 350	90 to 150
0000	336,400 C.M.	350 to 400	150 to 215

(These values are based on thermal limitations of the wire or cable, and are only approximate.)

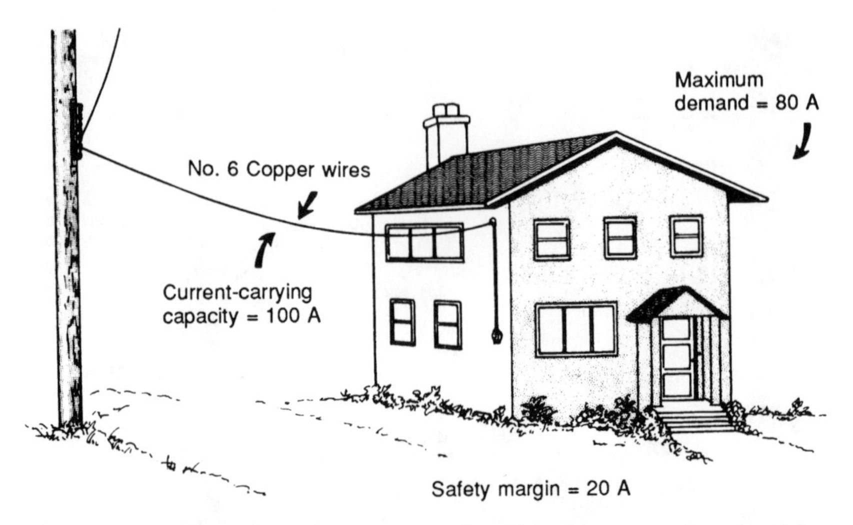

Figure 7-11. Choosing the proper wire size for a service provides a margin of safety.

where only the actual consumption of electricity is measured. Although it would be desirable to measure the maximum demand in these cases, the expense of the additional equipment is not warranted.

Two types of watt-hour meters (see Figure 7-12) are manufactured: the so-called A-base type and the "socket" or S type.

A-Base Watt-Hour Meter

The A-base has the terminals of the meter terminating in a block at the base of the meter and is essentially an indoor type of meter. A separate enclosing box is required when it is installed outdoors. The meter is mounted flush above a meter box through which the service conductors pass and in which the necessary connections are made. This box is sealed to prevent energy diversion.

The Socket Watt-Hour Meter

The socket or type S meter shown in Figure 7-13 has its terminals connected to bayonets or plugs in the back of the meter which fit into receptacles acting as terminals for the service conductors. No separate connections to the meter are necessary. The meter sets flush on the socket and is locked in place with a ring which is sealed. This type meter, which may be mounted either outdoors or indoors, reduces meter

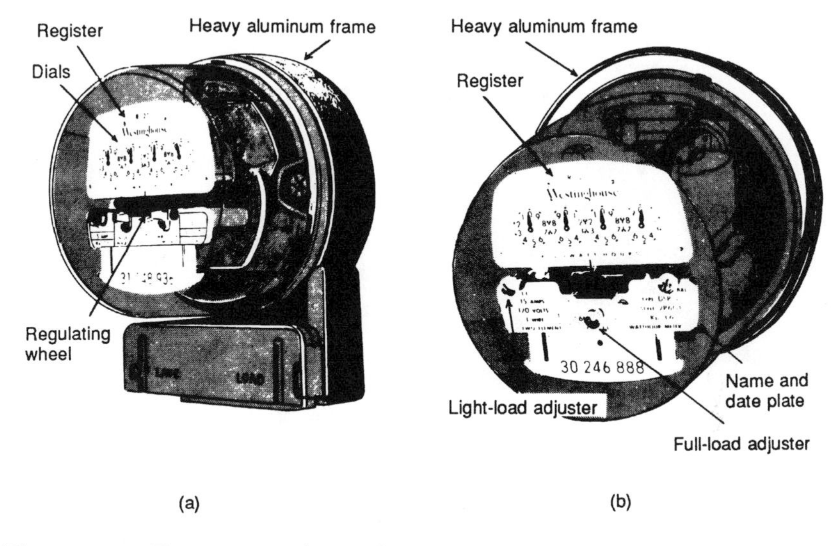

Figure 7-12. Two types of watt-hour meters. (a) A-base and (b) socket S-type.

installation to simply plugging in the instrument and sealing the ring. Outdoor meters are coming into wider use as they facilitate reading, testing, and replacement without disturbing the consumer. The socket type meter can be quickly replaced, with only a very brief interruption of service.

Meter Registers

An important part of the meter is the revolution counter from which the consumers' bills are determined. These registers are made in two types: the conventional dial type, and the cyclometer type.

In the dial type, the revolutions are counted on four dials (see Figure 7-14): the one at the extreme right registers units of electricity consumed. The second dial from the right is actuated by the one at the extreme right through a set of gears having a ratio of 10:1, so that it makes 1 complete turn for every 10 turns made by the one on the extreme right. Similarly, the third dial from the right is actuated by the dial second from the right through another 10:1 ratio set of gears—so that it makes 1 turn for every 10 made by the dial second from the right and for every 100 made by the dial at the extreme right. Again, the left hand dial or the dial fourth from the right is actuated by the dial third from the right through another 10: 1 ratio set of gears—so that it makes 1 turn

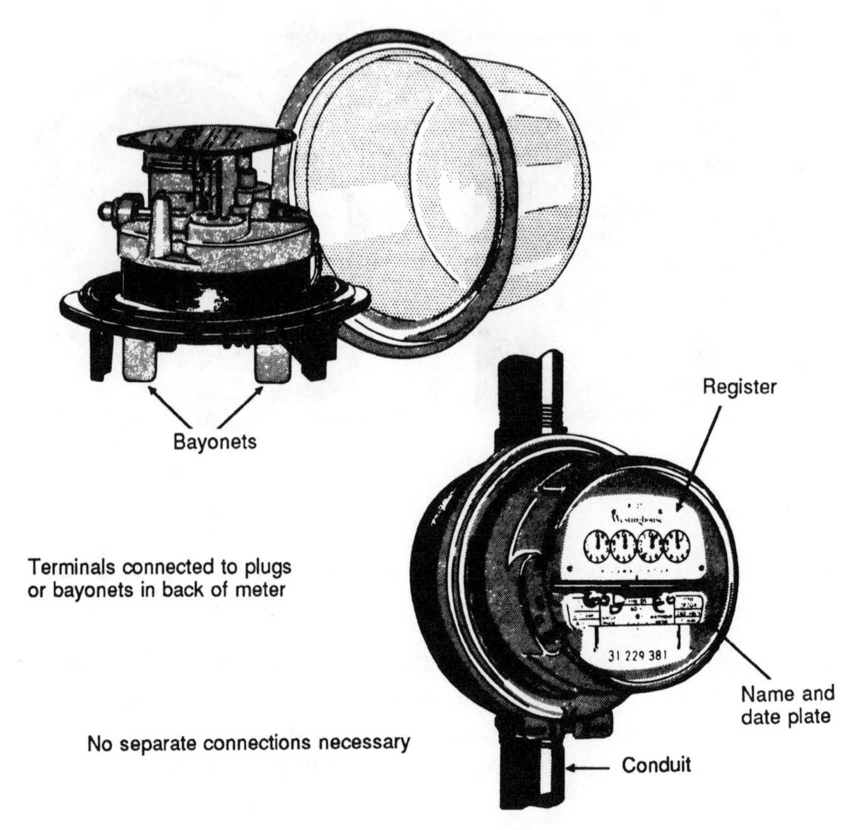

Figure 7-13. The socket watt-hour meter is easy to install.

for every 10 made by the dial third from the right, 100 made by the dial second from the right, and 1000 made by the dial at the extreme right. Some registers have five dials where the left dial makes 1 turn for every 10,000 made by the dial at the extreme right.

Hence, the register measures the units of electricity passing through the meter: reading the right hand dial first, the number nearest the pointer is read; then the dial second from the right is read in the same manner giving the number of ten units measured; then the dial third from the right in the same manner giving the number of hundred units measured; and, finally the first dial or fourth from the right is read in the same manner giving the number of thousand units measured.

In the cyclometer type register (shown in Figure 7-15), although the action is the same as in the conventional type, the dial faces are replaced

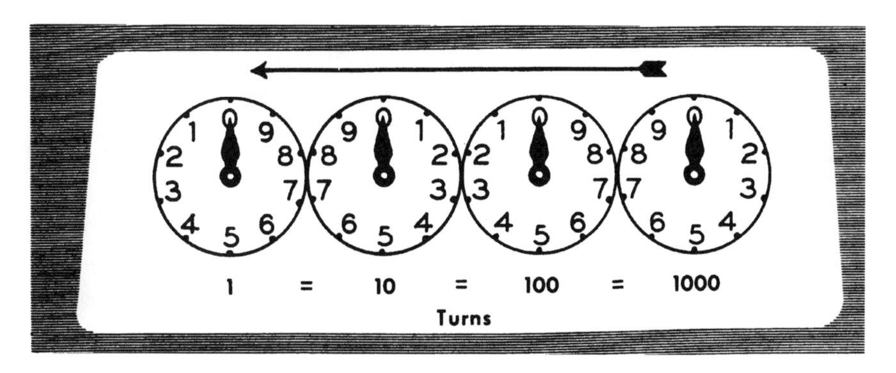

Figure 7-14. The dial register.

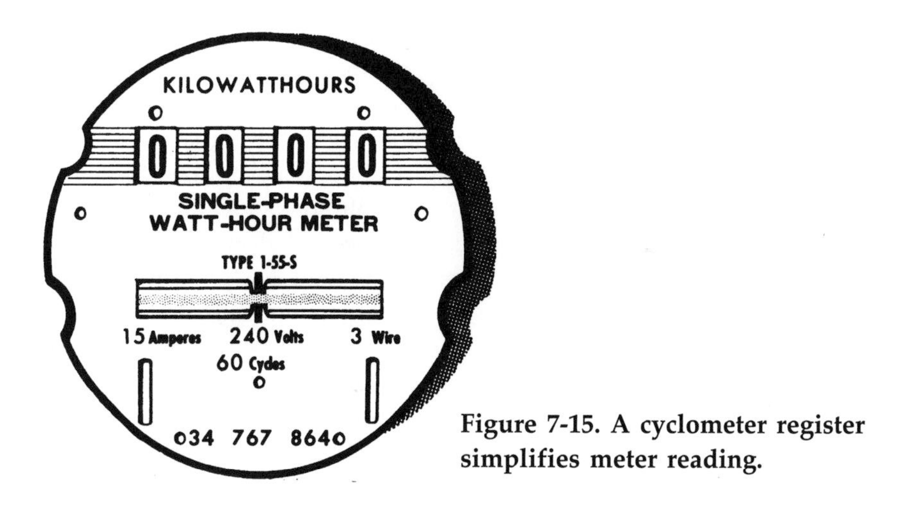

Figure 7-15. A cyclometer register simplifies meter reading.

by wheels which indicate numbers directly, making possible the reading of the meter directly, from left to right. This simplifies reading of the meter and takes some of the mystery out of the meter for the average consumer.

Billing is accomplished by taking the difference in readings of the meter at two successive times at an interval of usually one month. The difference in readings indicates the amount of electricity (in kilowatt hours) used by the consumer in that period. This is multiplied by the appropriate rate and the bill is rendered to the consumer (see Figure 7-16).

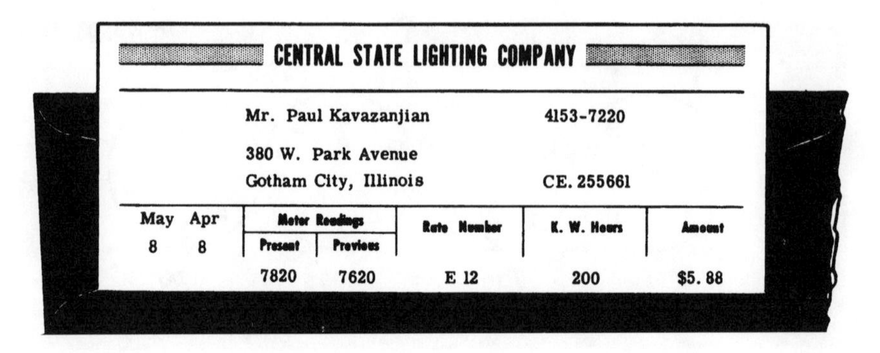

CENTRAL STATE LIGHTING COMPANY

Mr. Paul Kavazanjian 4153-7220

380 W. Park Avenue

Gotham City, Illinois CE. 255661

May	Apr	Meter Readings		Rate Number	K. W. Hours	Amount
8	8	Present	Previous			
		7820	7620	E 12	200	$5. 88

Figure 7-16. A typical monthly bill rendered to a residential consumer.

Watt-Hour Demand Meters

For large consumers, it is profitable to measure the actual maximum demand. This is done by adding another dial and set of gears to the watt-hour meter, converting it to a demand meter. There are two hands on this dial, one of which is connected to the actuating mechanism and the other just "floats," being pushed along by the first hand. The first hand measures the number of revolutions made in a standard period of time—15 minutes (in special instances a half-hour or an hour). At the end of this period, another mechanism operates to return the first hand to zero and the process starts all over. In the meantime, the second hand having been pushed by the first hand, remains at the last place before the first hand is returned to zero. Hence, the second hand measures the maximum demand of the consumer. A meter equipped with only this second type of gear set and dial is known as a *demand meter* (see Figure 7-17). Where a meter is equipped with both sets of gears and dials, it is known as a *watt-hour demand meter.*

In reading this type of meter for billing purposes, the regular four dials are read for the kilowatt hour consumption and then the floating hand is read for the kilowatt demand. Through a lever protruding through the case, this floating hand is returned to zero each time the meter is read, preparing it to register the maximum demand for the next metering period; this lever is locked to prevent tampering.

Since a record of this maximum demand value is lost as soon as the floating hand is returned to zero, a cumulative type of demand register has been developed, which translates this reading of the floating hand to another set of dials on another register similar to the one measuring consumption. The demand reading on this register stays put, being

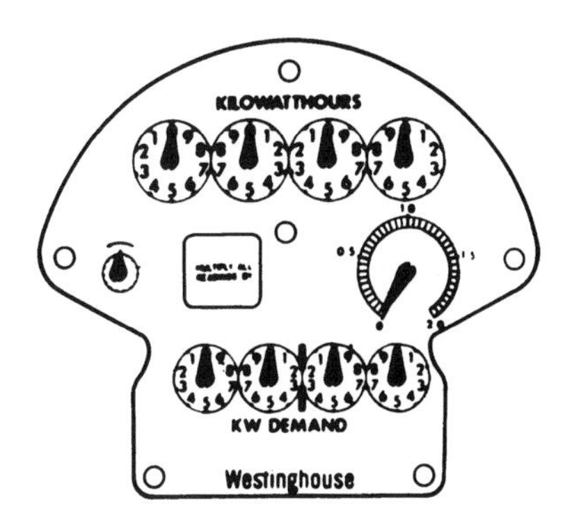

Figure 7-17. The register of a demand meter—for large consumers.

available for checking, until the meter is next read, when the lever or plunger which returns the "floating" hand to zero causes the new maximum demand to be added to the original reading: the new reading will then be greater than the previous one by the current period's maximum demand.

SERVICE EQUIPMENT

When the meter is placed indoors, the service conductors coming into the consumers' premises terminate in a cut-out box. This is merely a point at which the service wires can be disconnected from the house wiring. It may be a switch and each conductor may or may not be fused. From this point, the wires extend to another box associated with the meter. This second box holds the terminals for the meter and a special switch which permits testing the meter without interrupting service. Where only one meter of small size is installed, the service cut-out box, the meter terminal, and test box may be combined in one cabinet. In any case, these boxes are all locked with a seal which must be broken when they are opened. This is done to prevent the theft of electricity by by-passing or tampering with the meter. In more recent installations, the positions of these two boxes are reversed, the box associated with the meter comes first and the switch box next (see Figure 7-18).

The equipment to this point, with the exception of the meter, may or may not be owned by the consumer. In all cases, however, the utility company owns the meter.

The service from this point is then connected to a distribution panel which supplies the circuit to the various parts of the consumer's premises. These circuits are usually protected at the distribution panel [see Figure 7-19(a)] so that in the event of an overload, or an accidental short circuit in one of the circuits, only a fuse or breaker will operate and service to the remaining parts of the consumers' premises will not be interrupted. Most installations today provide that this panel may be contained in the same box as the switch.

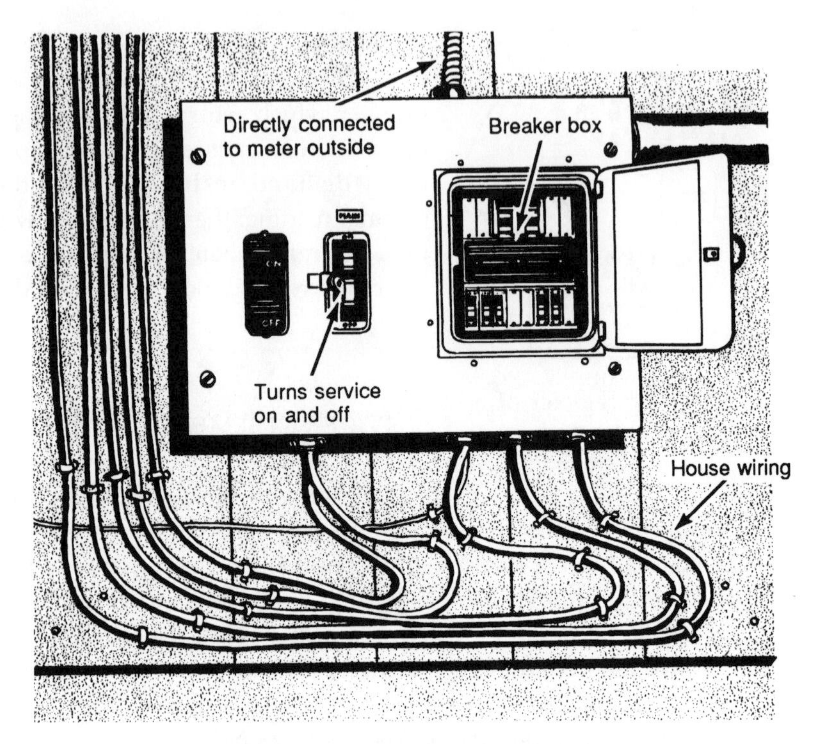

Figure 7-18. An enclosed switch box.

When the meter is placed outdoors, there is no need for the box providing a point to connect or test the meter. The remaining distribution panel is still required and is usually installed inside the consumer's premises.

For safety purposes, one conductor is connected to ground, usually to a driven rod or pipe or other metallic structure [see Figure 7-19(b)]. This is done so that if an accidental contact of an energized conductor takes place, the energy is conducted to the ground and possibility of electric shock reduced. For this reason, a fuse is never placed in the circuit of this "neutral" or ground conductor.

REMOTE METER READING AND DEMAND CONTROL

Electronic developments that have made e-mail (and the Internet) inexpensive and universal means of communication have also made practical the remote reading of consumer's meters. Periodic inquiry

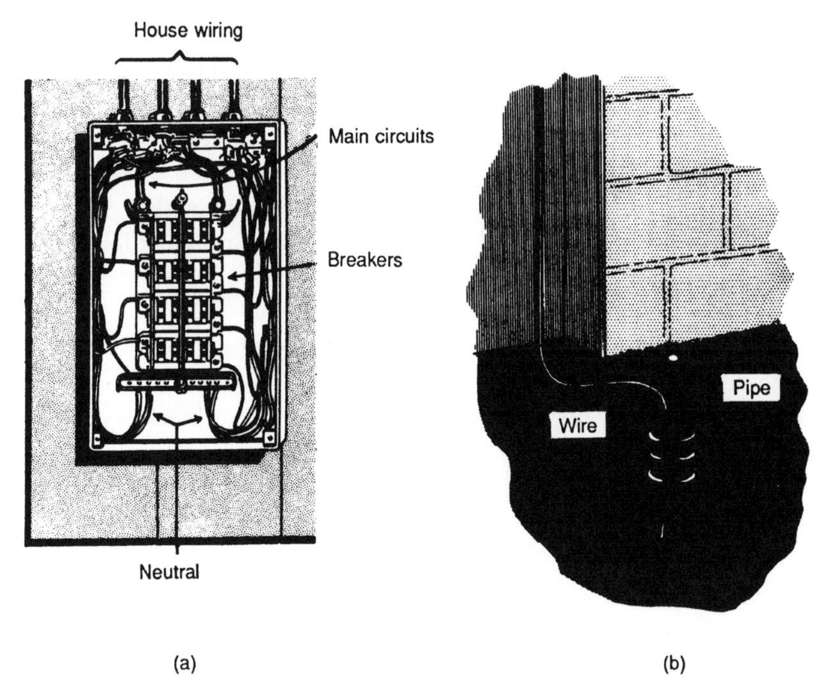

Figure 7-19. (a) An exposed switch box and (b) grounding a meter wire is a safety measure.

automatically sent to each consumer identifies their meter and records the dial consumption and other data, transmitting them to the computer center where they may be automatically processed, producing the bill sent to the consumer.

In some cases, usually commercial or industrial consumers, where it is desired to hold down their demands by arranging their individual loads not to coincide, and where practical to be scheduled for off peak hours (usually evening and early morning hours), the same means of communication is used to operate relays and switches to accomplish this purpose, often employing the same meter reading facilities.

REVIEW QUESTIONS

1. What is meant by *maximum demand*?
2. Define *load-factor; diversity factor. How* are they determined?

3. What is an electric *service*?
4. What data about the load is used to determine the size of service wires or cables?
5. Describe an overhead service and its installation; also, an underground service and its installation.
6. What is the function of the electric watt-hour meter?
7. What is the difference between the conventional dial and the cyclometer types of registers?
8. What is the function of the demand meter?
9. What is the purpose of the cumulative register associated with the demand meter?
10. What other equipment complete the consumer's service requirements?

Chapter 8

Substations

FUNCTIONS OF A SUBSTATION

Substations serve as sources of energy supply for the local areas of distribution in which they are located. Their main functions as shown in Figure 8-1 are to receive energy transmitted at high voltage from the generating stations, reduce the voltage to a value appropriate for local use, and provide facilities for switching. Substations have some additional functions. They provide points where safety devices may be installed to disconnect circuits or equipment in the event of trouble. Voltage on the outgoing distribution feeders can be regulated at a substation. A substation is a convenient place to make measurements to check the operation of various parts of the system. Street lighting equipment as well as on-and-off controls for street lights can be installed in a substation though this is a diminishing function.

Some substations are simply switching stations where different connections can be made between various transmission lines.

TYPES OF SUBSTATIONS

Some substations are entirely enclosed in buildings as shown in Figure 8-2(a), while others are built entirely in the open [see Figure 8-2(b)], for example, unit substations. In this type, all the equipment is assembled into one metal clad unit usually enclosed by a fence. Other substations have step-down transformers, high-voltage switches and oil circuit breakers, and lightning arresters located just outside the substation building within which are located the distribution and street lighting facilities.

Sites for substations are generally selected so that the stations will be as near as possible to the load center of the distribution areas which they are intended to serve. Availability of land, cost, local zoning laws,

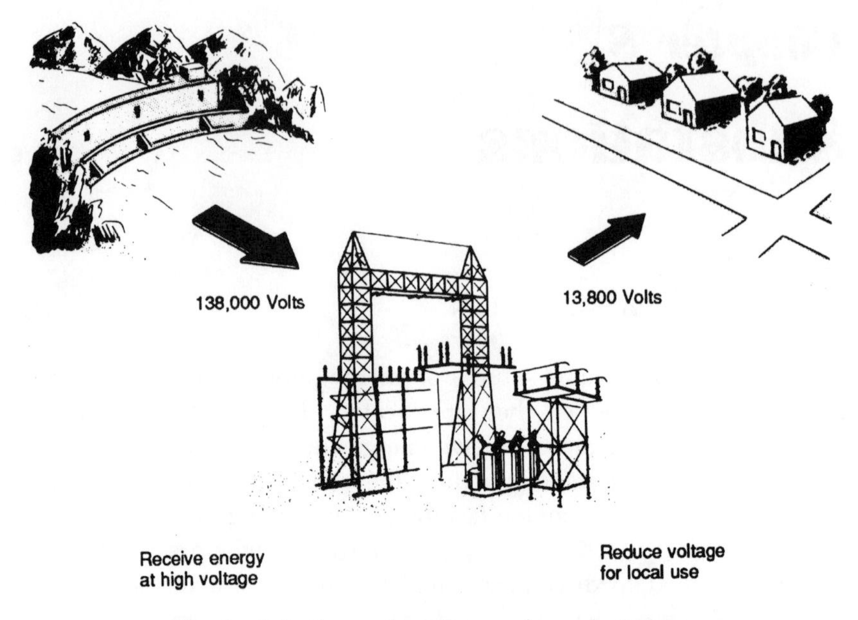

Figure 8-1. Some functions of a substation.

future load growth, and taxes are just a few of the many factors which must be considered before a site is ultimately chosen.

Substations may have an operator in attendance (see Figure 8-3) part or all of the day, or they may be entirely unattended. In unattended substations, all equipment function automatically, or may be operated by remote control from an attended substation or from a control center.

UNIT SUBSTATIONS

Utility companies at one time concentrated many feeders into a few big, attended distribution substations. Each one of these substations had to be individually engineered and there was little standardization.

However, the unit substation offers considerable economy in investment and installation. These factory-built, metal-enclosed units which usually supply only a few feeders give the system a high degree of standardization (see Figure 8-4).

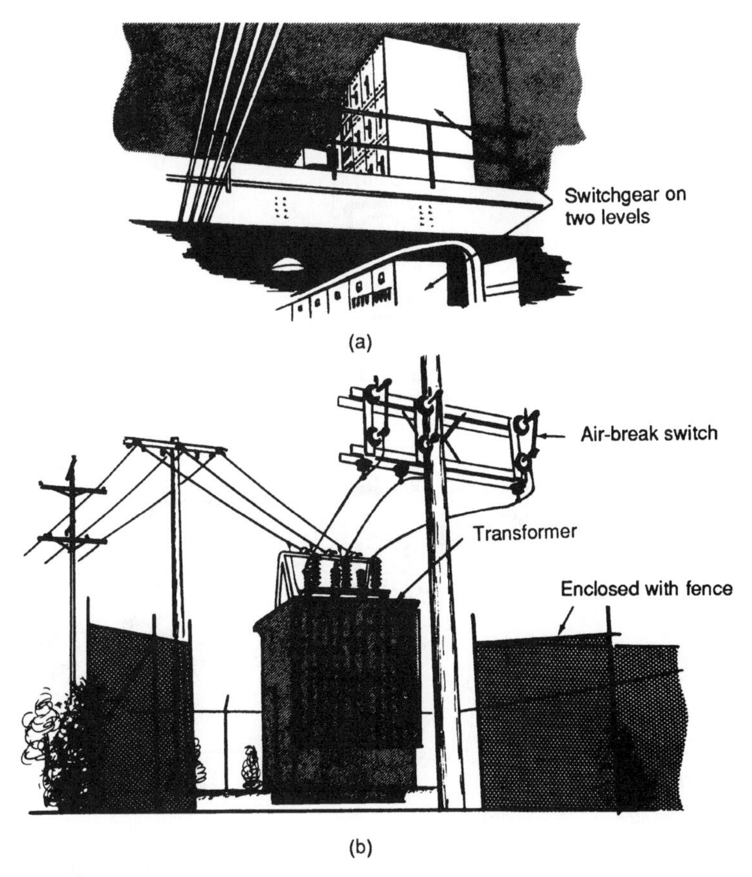

Figure 8-2. Substation types: (a) indoor and (b) outdoor.

METAL-CLAD SWITCHGEAR

In some types of modern substations, the circuit breakers, disconnecting switches, buses, measuring instruments, and relays associated with the distribution circuits are enclosed in metal cabinets (see Figure 8-5), one for each circuit, providing fire and explosion protection between cabinets. The number of cabinets so mounted together vary with the number of circuits involved. This makes a neater appearance, and lessens the expense necessary to protect this equipment from the weather. Also, these compartments are all prewired at the factory and result in economy in such installations.

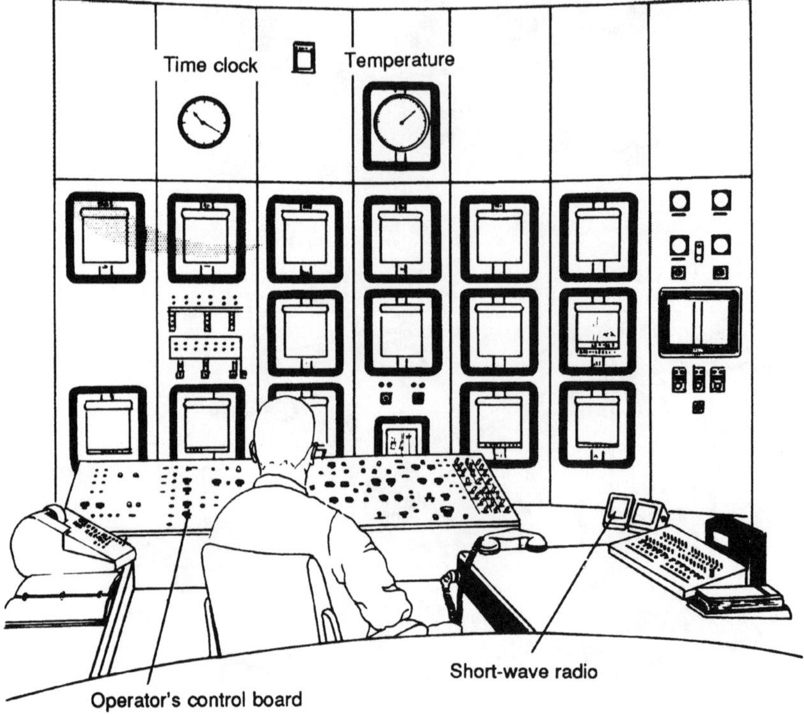

Figure 8-3. A control center may operate several substations by remote control.

SUBSTATION EQUIPMENT

Transformers

Transformers for changing the voltage from that of the incoming supply to that of the outgoing distribution feeders operate on the same principle as do distribution transformers and are constructed in substantially the same manner. They are usually filled with oil and are cooled either by the surrounding atmosphere or by air blasts obtained from fans trained on them. Some also circulate the oil for additional cooling.

Because of the high voltages imposed on the incoming side, there is an elaborate electrical connection going through the cover which is called a *bushing*.

Figure 8-4. A unit substation is conveniently installed.

The supply circuit is connected to the terminals of the primary winding and the outgoing distribution feeders connected to the terminals of the secondary winding. Substation transformers (see Figure 8-6) are generally used to reduce the voltage to a lower value for distribution and utilization. They can also be used in some situations to step up voltage.

Regulators

The functioning of a regulator was previously discussed in the section on line equipment. However, mention of its basic principle is repeated here.

A regulator, shown in Figure 8-7, is really a transformer with a variable ratio. When the outgoing voltage becomes too high or too low for any reason, this apparatus automatically adjusts the ratio of transformation to bring the voltage back to the predetermined value. The adjustment in ratio is accomplished by tapping the windings, varying the ratio by connecting to the several taps. The unit is filled with oil and is cooled

(a)

(b)

Figure 8-5. Metal-clad switchgear. (a) Installing a circuit breaker in an indoor substation and (b) adjusting a control in an outdoor substation.

much in the same manner as a transformer. A panel mounted in front of the regulator contains the relays and the other equipment which control the operation of the regulator.

Circuit Breakers

Oil circuit breakers (Figure 8-8) are used to interrupt circuits while current is flowing through them. The making and breaking of contacts are done under oil. As explained previously, the oil serves to quench the arc when the circuit is opened. The operation of the breaker is very rapid when opening. As with the transformer, the high voltage connections are made through bushings. Circuit breakers of this type are usually arranged for remote electrical control from a suitably located switchboard.

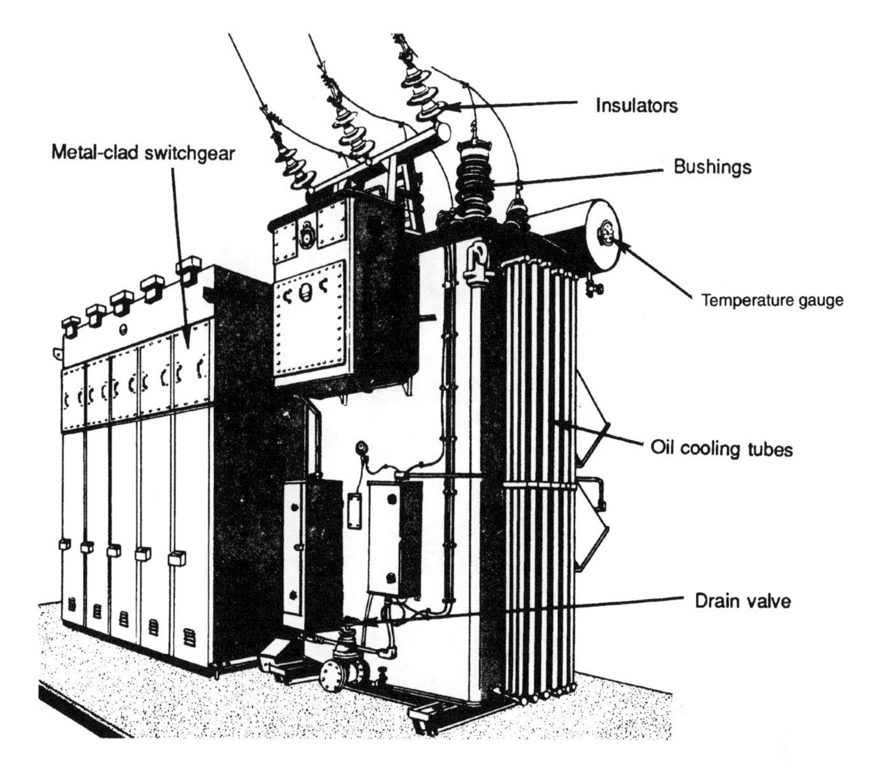

Figure 8-6. A substation transformer.

Some circuit breakers have no oil, but put out the arc by a blast of compressed air. These are called *air circuit breakers.* Another type encloses the contacts in a vacuum or a gas (sulfur hexafluoride, SF_6) which interrupts the conductive path of the arc.

Air-Break and Disconnect Switches

The operation of switches was discussed in detail in Chap. 4. Some are mounted on an outdoor steel structure called a *rack* (see Figure 8-9), while some may be mounted indoors on the switchboard panels. They are usually installed on both sides of a piece of equipment, to effectively de-energize it for maintenance.

The Switchboard

This is usually a panel or group of panels made of some insulating material on the front of which are mounted the various meters, relays, controls, and indicators for the proper operation of transform-

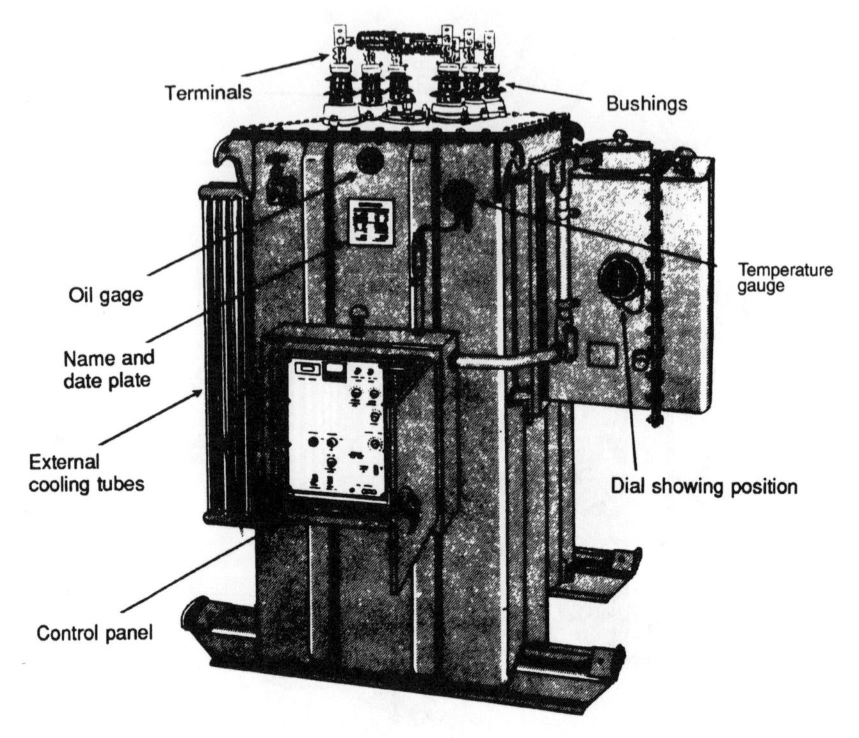

Figure 8-7. A substation voltage regulator.

ers, circuit breakers, and other equipment located in the substation. The switchboard shown in Figure 8-10 contains all the equipment necessary for controlling, protecting, and recording what goes on in a substation.

Measuring Instruments

Various instruments are mounted on the switchboard to indicate to the operator what voltage is impressed on the various incoming and outgoing feeders as well as on various pieces of equipment, and the current and power that flows through this equipment (see Figure 8-11). The instrument that measures voltage is called a *voltmeter;* that which measures current is known as an *ammeter;* and that which measures power is called a *wattmeter.* These instruments may be either indicating or recording. An indicating instrument indicates the quantity at a particular moment, usually by position of the needle on the scale. A recording instrument not only indicates the present value of

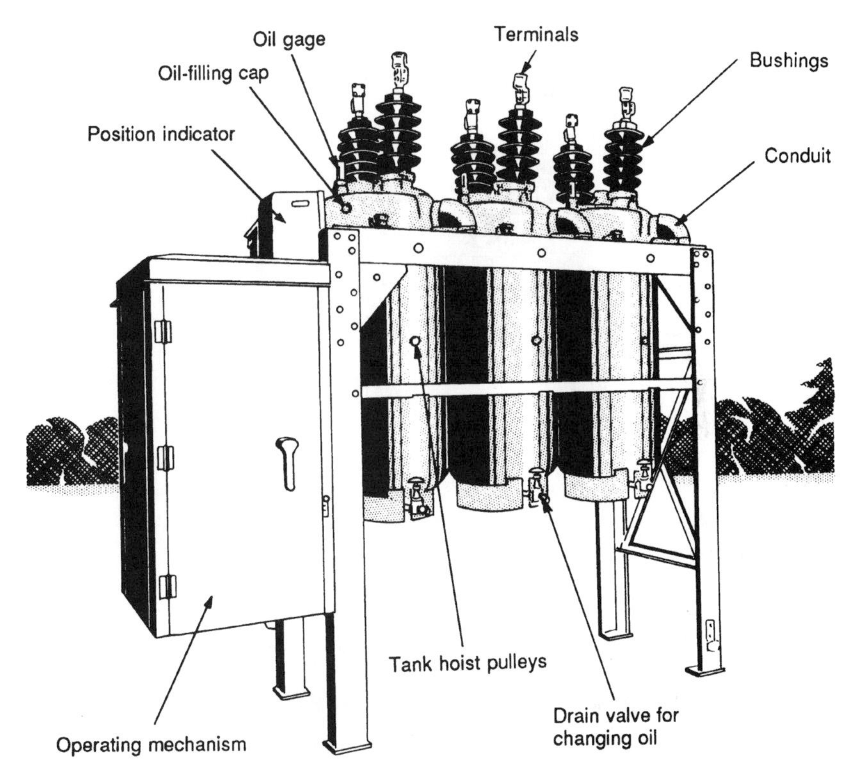

Figure 8-8. Substation oil-circuit breaker.

the quantity, but also records past values over a period of time. This is usually done by tracing a line on a chart or graph.

Relays

A relay is a low-powered device used to activate a high-powered device. In a transmission or distribution system, it is the job of relays to give the tripping commands to the right circuit breakers.

Relays (Figure 8-12) protect the feeders and the equipment from damage in the event of fault. In effect, these relays are measuring instruments, similar to those in Figure 8-11, but equipped with auxiliary contacts which operate when the quantities flowing through them exceed or go below some predetermined value. When these contacts operate, they in turn actuate mechanisms which usually operate switches or circuit breakers, or in the case of the regulator, operate the motor to restore voltage to the desired level. Modern relays employ electronic circuitry.

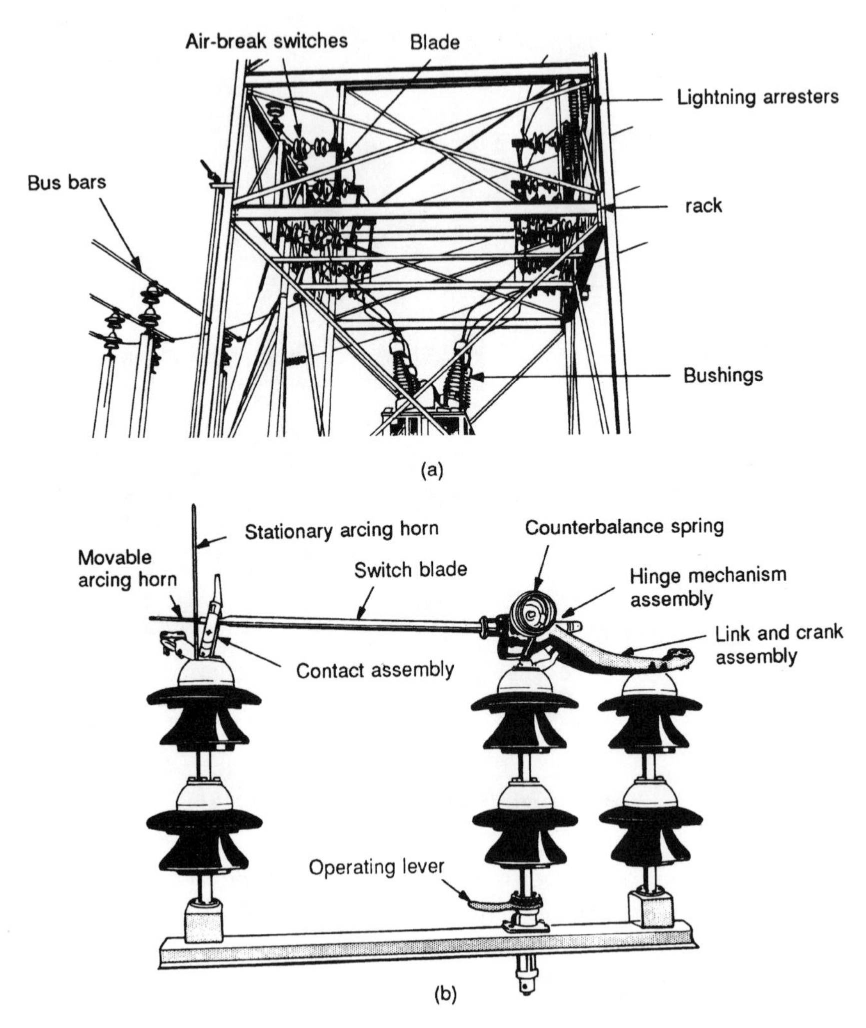

Figure 8-9. Air-break switches. (a) Mounted on a substation rack. (b) Detailed drawing of an air-break switch.

Bus Bars

Bus bar (bus, for short) is a term used for a main bar or conductor carrying an electric current to which many connections may be made.

Buses are merely convenient means of connecting switches and other equipment into various arrangements. The usual arrangement of connections in most substations permits working on almost any piece of equipment without interruption to incoming or outgoing feeders.

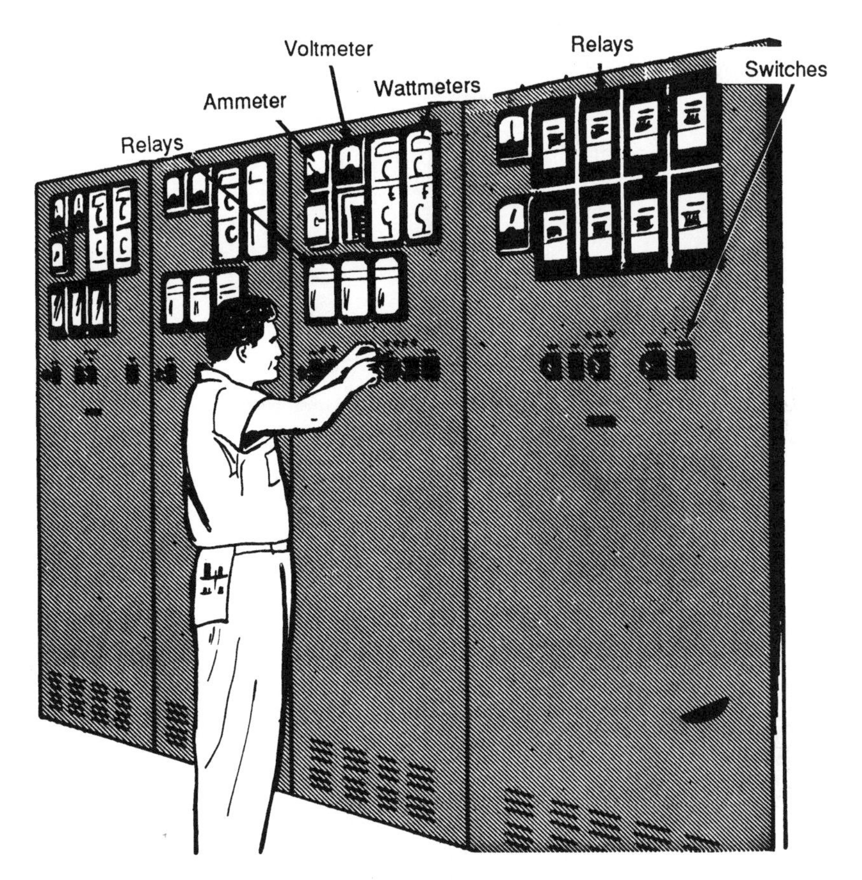

Figure 8-10. Substation switchboard.

Some of the arrangements provide two buses to which the incoming or outgoing feeders and the principal equipment may be connected. One bus is usually called the *main* bus and the other *auxiliary* or *transfer* bus. The main bus may have a more elaborate system of instruments, relays, and so on, associated with it. The switches which permit feeders or equipment to be connected to one bus or the other are usually called *selector* or *transfer* switches. Bus bars come in a variety of sizes and shapes, shown in Figure 8-13.

Substation Layout

The equipment and apparatus described may be connected together as shown in the typical layout of Figure 8-14(a). Description of symbols used is shown in Figure 8-14(b).

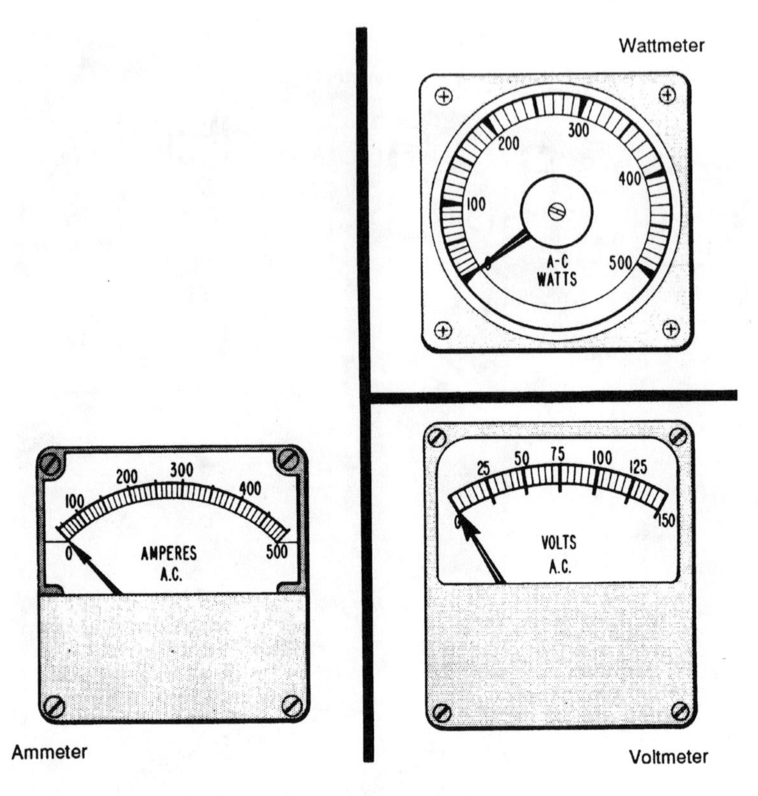

Figure 8-11. Indicating instruments used in a substation.

POLYMER (PLASTIC) INSULATION

In general, any form of porcelain insulation may be replaced with polymer. Their electrical characteristics are about the same. Mechanically they are also about the same, except porcelain must be used under compression while polymer may be used both in tension and compression. Porcelain presently has an advantage in shedding rain, snow and ice and in its cleansing effects; polymer may allow dirt, salt and other pollutants to linger on its surface encouraging flashover, something that can be overcome by adding additional polymer surfaces. Polymer's greatest advantage, however, is its comparatively light weight making for labor savings; and essentially no breakage in handling. For detailed comparison of the two insulations, refer to Appendix A, Insulation: Porcelain vs. Polymer.

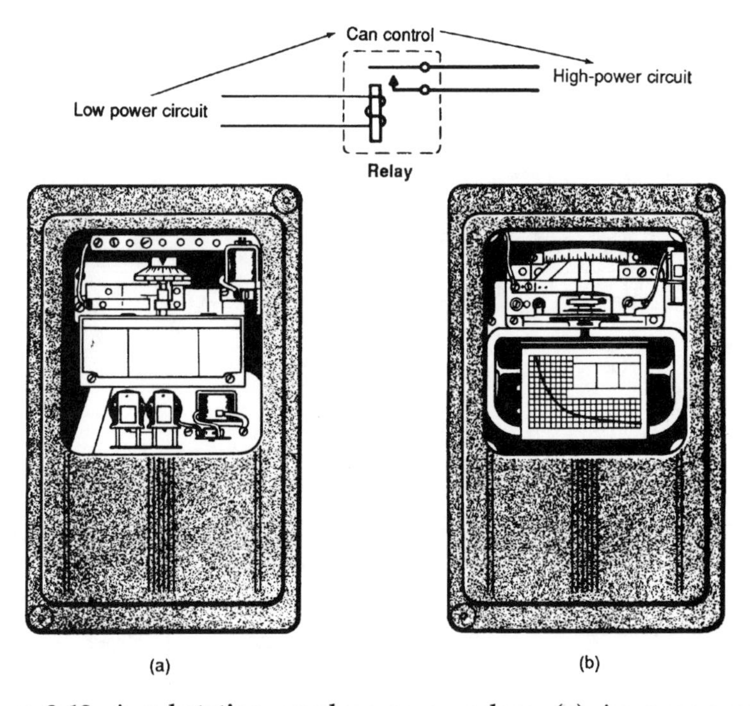

Figure 8-12. A substation employs many relays. (a) An over-current relay and (b) an over-voltage relay.

REVIEW QUESTIONS

1. What is the main function of a substation? What are some of its other functions?
2. What important factor influences the location of substations? What other factors also receive consideration?
3. What is the advantage of the unit substation?
4. Where and why is metal-clad switchgear used?
5. What is the function of a regulator?
6. What do substation transformers do?
7. What does a circuit breaker do?
8. What are the functions of air-break switches and disconnects?
9. What is the value of a relay? How does it function in a substation?
10. How may bus bars vary?

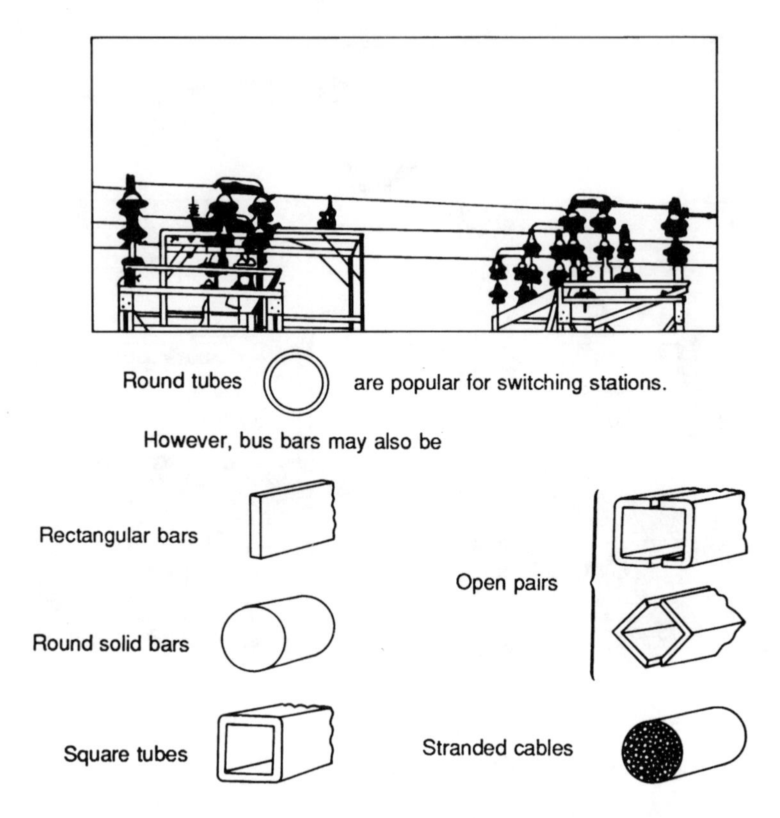

Figure 8-13. Typical types of bus bars in a substation.

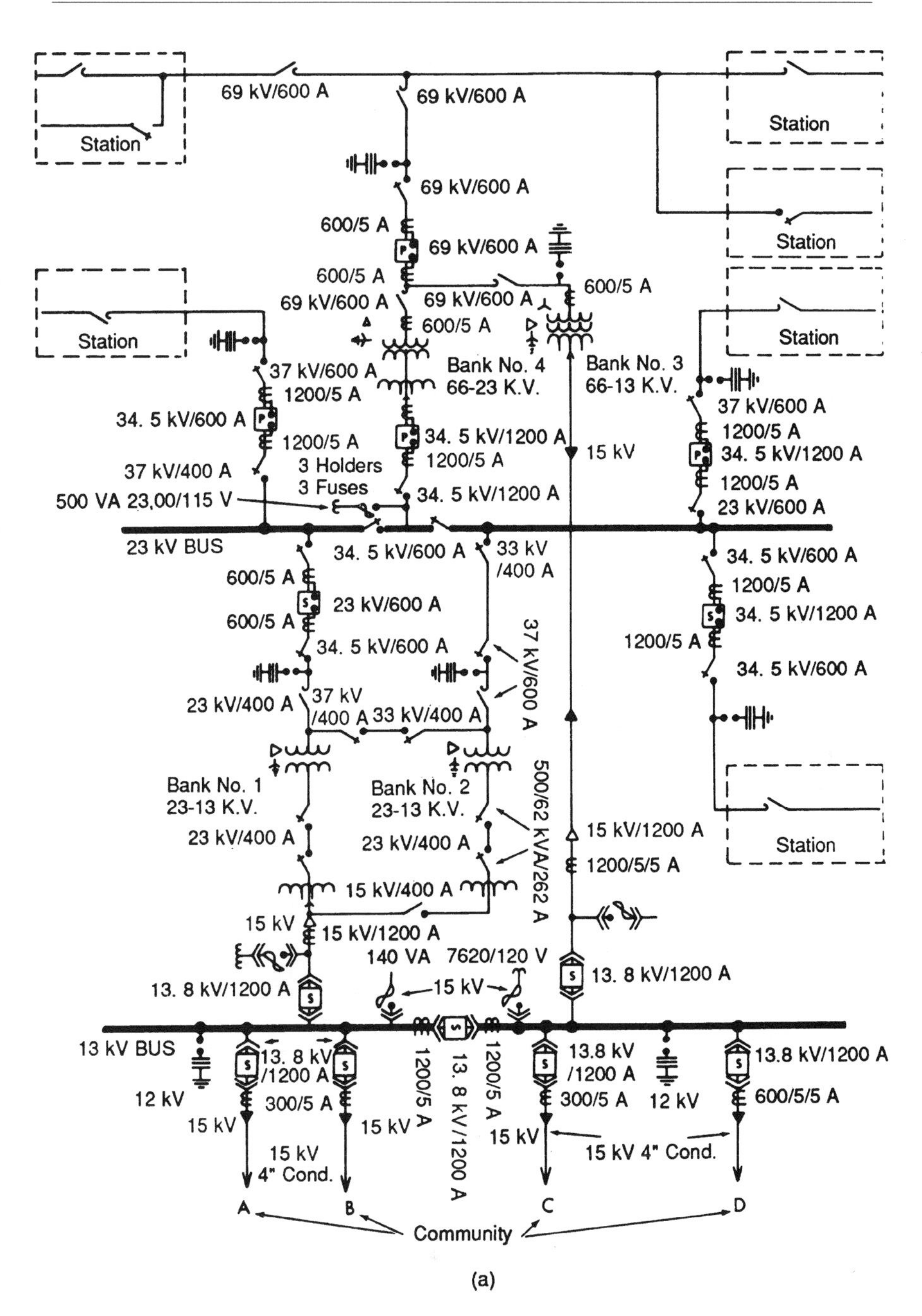

Figure 8-14. (a) Typical arrangement of a substation.

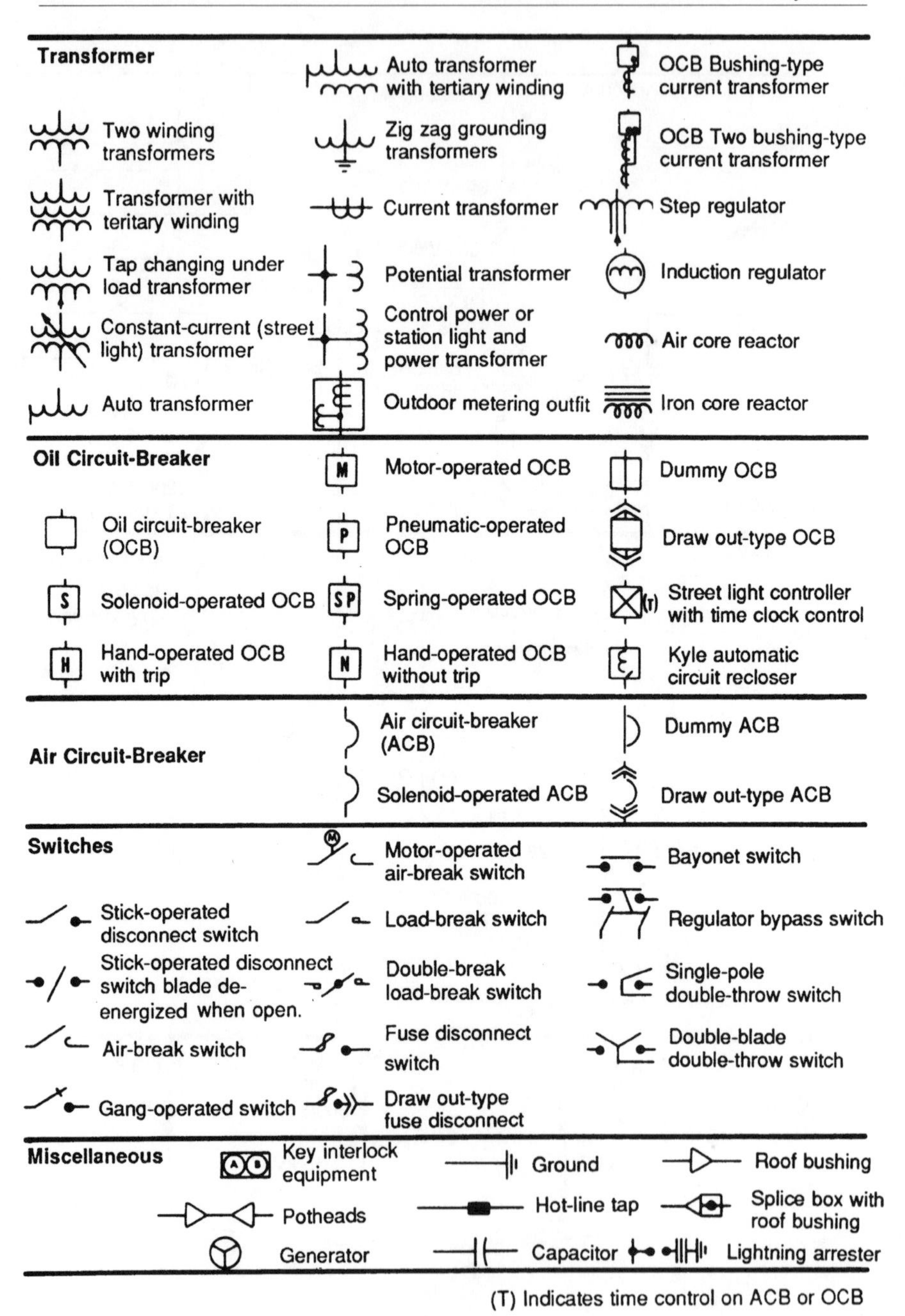

Figure 8-14. (b) Standard symbols for equipment.

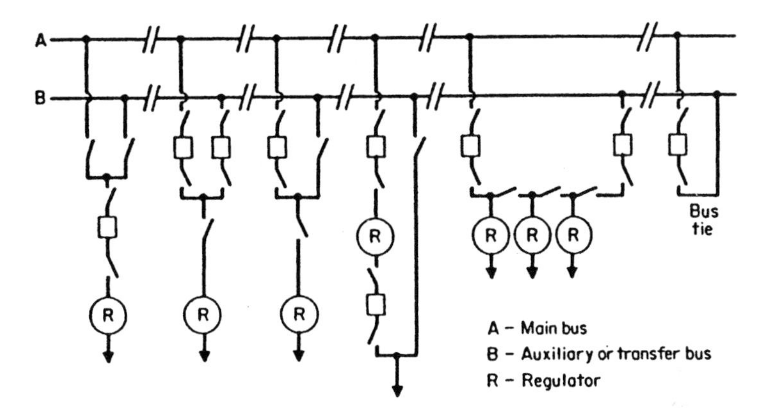

Figure 8-15. Arrangement of Distribution feeder Buses at Substations. (*Courtesy Westinghouse electric Co.*)

Chapter 9

Distribution Circuits, Cogeneration and Distributed Generation

Distribution circuits generally consist of two parts Figure 9-1: the primary circuit operating at a relatively high pressure or voltage that carries the electric supply to the area here it is to be used. The secondary circuit receives the supply from the primary through transformers that reduce its pressure or voltage to values low enough to deliver the product safely to consumers Typical primary circuit voltages are 2400, 4160, 7620, 13800 and 23000 volts; secondary circuit voltages approximate 120 and 240 volts.

THE PRIMARY CIRCUIT

Should a fault occur on the circuit shown, Figure 9-2a, a large current will flow to the fault that, if permitted to flow, will eventually burn the conductor apart, 'clearing' the fault, but with hazards to the public. Relays at the substation, however, will sense this large 'fault' current and operate to open the circuit breaker deenergizing the entire

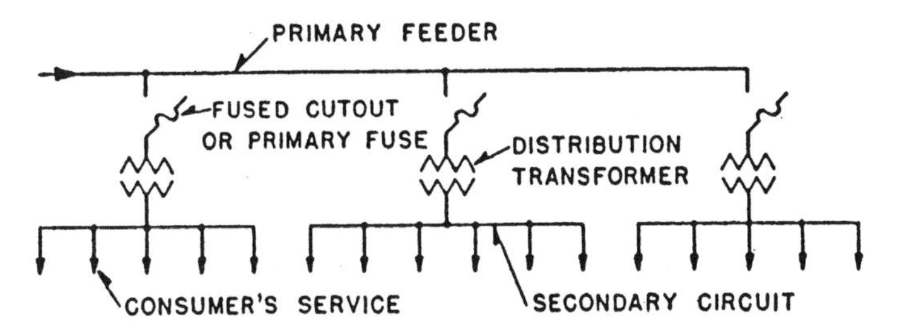

Figure 9-1. Schematic diagram showing connections of transformers between primary feeder and secondary circuits in a distribution system.

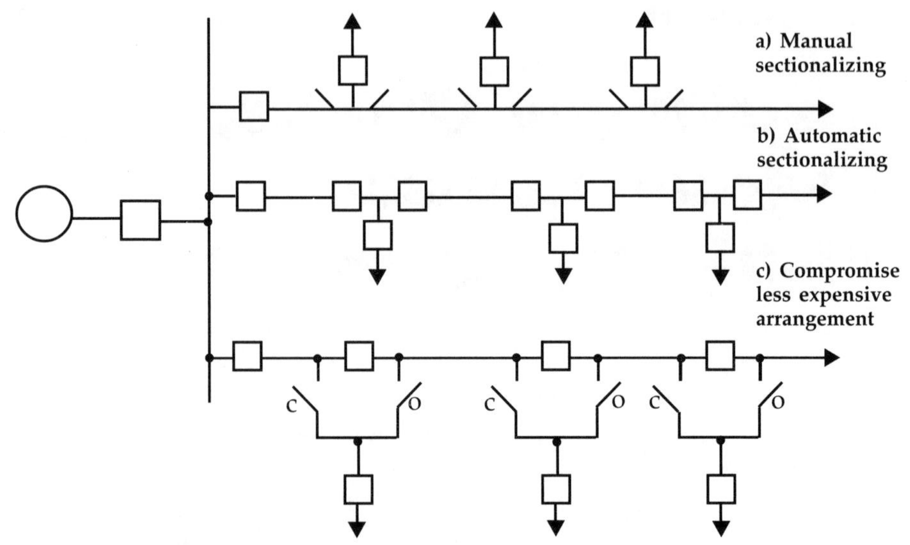

Figure 9-2. a-Schematic diagram of radial type circuits, showing several methods of sectionalizing.

circuit leaving consumers without power until repairs are made. This is a basic circuit known as a 'radial' circuit.

LOOP CIRCUITS

To shorten the time of outage, the circuit may have a number of sectionalizing switches as shown in Figure 9-2b, allowing parts to be connected to other unfaulted circuits, restoring service to some consumers. If the circuit is made into a loop, Figure 9-3a, with both ends connected to its source through one or two circuit breakers, the fault can be confined to the one section on which the fault occurred by opening the sectionalizing switches on both ends of the faulted section and closing the one or two breakers at the substation. In the closed loop type of circuit, the fault current flowing to the fault is divided into two parts, flowing in each side of the fault. This may sometime be too small to achieve positive operation of the relays to open the breaker or breakers. If the loop now be deliberately open at some point, the entire fault current will flow through the branch between the fault and the substation breaker, assuring its positive operation. The two ends of the circuit at the open point are connected together through a circuit breaker set to close

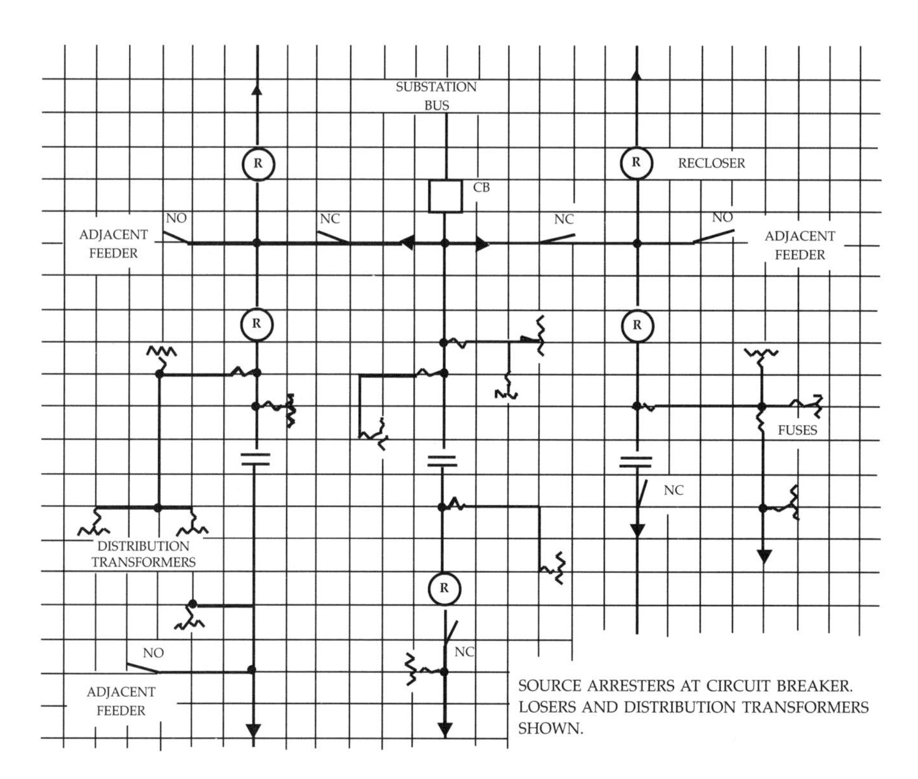

Figure 9-2. b-Radial primary feeder with protective and sectionalized devices.

automatically when one end is deenergized (or this may be accomplished manually with only a sectionalizing switch). All of the loop circuit will now be restored to service except the faulted section.

The sectionalizing switches may all be circuit breakers, Figure 9-3b, opening and closing automatically. Since circuit breakers are expensive compared to line switches, arrangements employing fewer breakers may be employed, as shown in Figure 9-3c, with a lessened degree of reliability for some consumers.

PRIMARY NETWORK

Where higher reliability is desired than from radial or loop circuits (such as commercial and industrial complexes where the units may not be spaced close together), the radial primary circuits are tied together

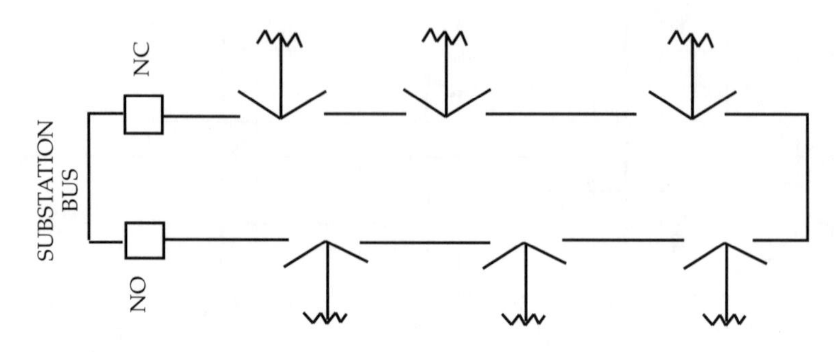

(a) Manual sectionalizing.

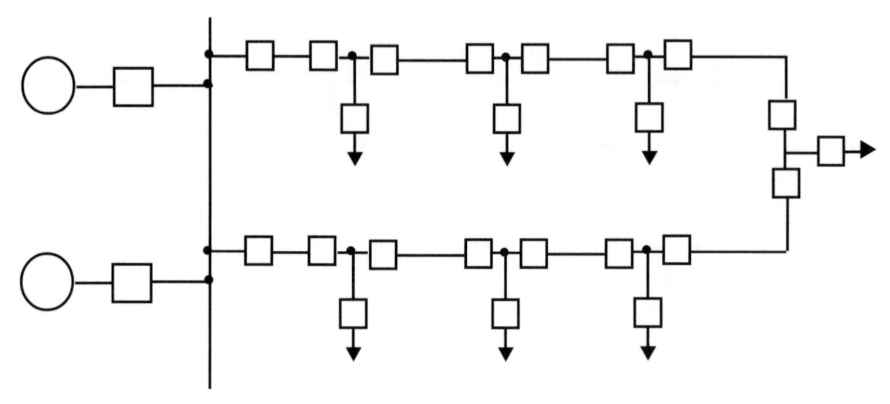

(b) Automatic sectionalizing.

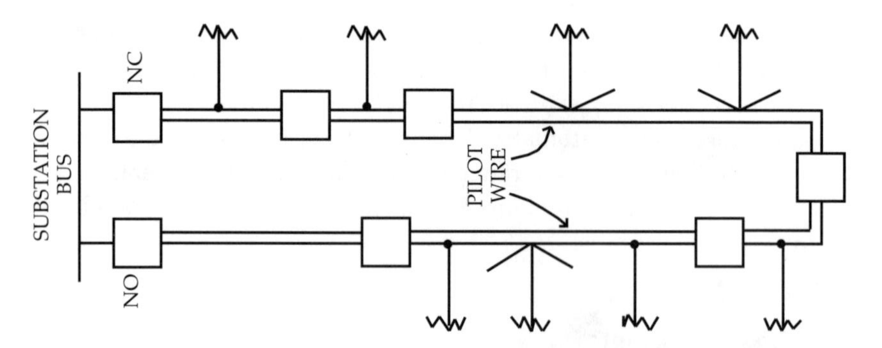

(c) Compromize sectionalizing less expensive.

Figure 9-3. Schematic diagrams—loop circuits (one circuit breaker arrangement shown in Figure 9-1a(c) also applies).

into a network. The network is supplied from a number of transformers or substations supplied in turn by subtransmission and transmission lines. Circuit breakers between the transformers and grid act as network protectors to protect the network from faults on the incoming high voltage lines. Figure 9-4. The high cost of operating such a system in providing for load growth and the high potential hazard involved with the enormous fault currents handled by the circuit breakers have led to their replacement by low voltage secondary 'spot' networks, though some primary networks may still be in operation.

SECONDARY CIRCUITS

The secondary circuits supplied through transformers from the primary systems are also affected by faults on their associated primary

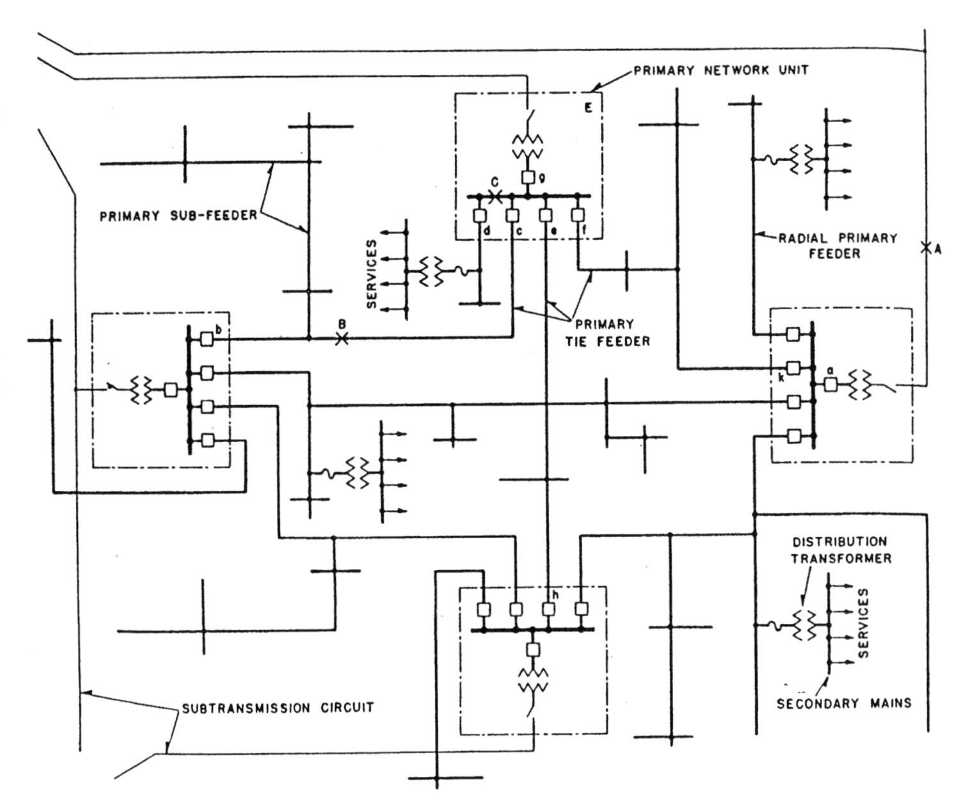

Figure 9-4. Typical primary-network arrangement using breakers.

supply, whether from a radial, closed or open loop circuit.

Where additional expense may be justified to achieve a higher degree of reliability, the secondary circuits of the several primary circuits may be connected together to form a network, Figure 9-5 so that service to any consumer will not be affected when a primary circuit, or any part of it, may become deenergized for whatever reason. The secondary network is connected through switches, called network protectors, through transformers connected to their primary group supply circuits. These switches, or protectors, are programmed to open when the current flowing through them reverses direction. This, so that current flow from the network may not supply fault current to a fault on any of the primary supply circuits.

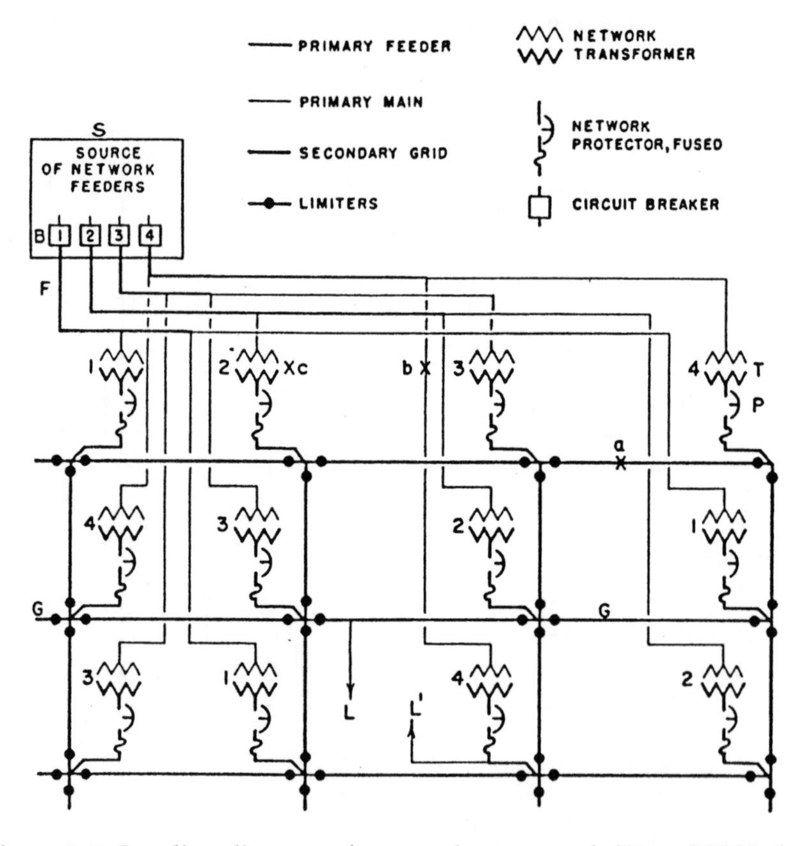

Figure 9-5. One-line diagram of a secondary network (*From EEI Underground Reference Book*)

The low voltage secondary network is also employed in serving large individual loads requiring a high degree of reliability; these are known as spot networks, Figure 9-6.

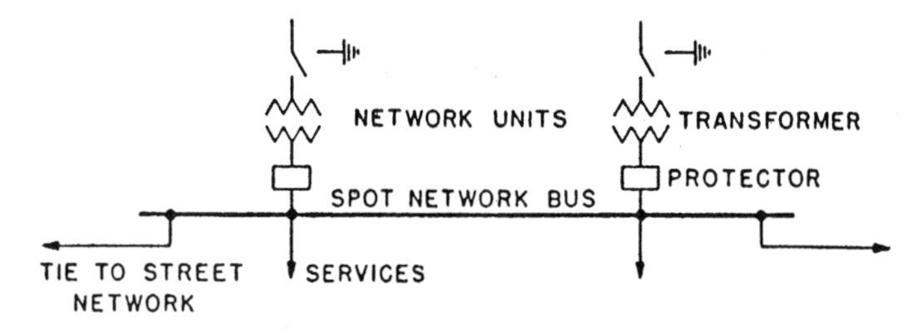

Figure 9-6. Typical spot network.

VOLTAGE REGULATION

The voltage variation at the consumer's service will depend on the current flowing from the source, the transmission and distribution systems right to the consumers service, The drop or difference in voltage from when the consumer's load is light on none to that when the consumers load is heaviest is referred to as voltage regulation, Figure 9-7.

COGENERATION

As a measure to conserve fuel, while at the same time helping to reduce the utility's financial investment in generating facilities (also sometimes difficult to locate because of environmental problems), generating facilities owned and operated by large consumers, may be connected directly to the utility's transmission and distribution lines, usually as a substation-type installation. Not only must necessary protective devices be installed, but they are generally under the supervision of the utility's coordinator or system operator. The cogenerating consumers sell their excess energy availability in the form of electricity to the utility; in some states, this sale is mandated by law.

Strict control of such installations is necessary as they may present a hazard to people working on the associated transmission and distribution lines. The line on which work is being done may be de-energized,

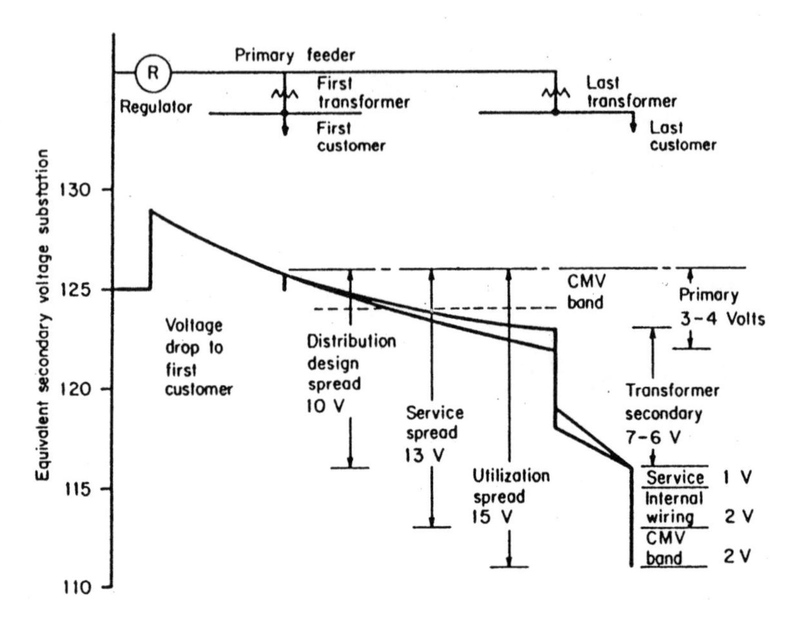

Figure 9-7. Allocation of voltage drops on distribution systems (*Courtesy Long Island Lighting Co.*)

but, unless the strict control is maintained, accidental re-energizing of the line may occur through the consumer closing the tying circuit breaker in error, or substandard protective equipment may fail. The line workers should take special precautions, usually by applying multiple grounds to the line on which they are to work. A one-line diagram of electric connections and protective devices is shown in Figure 9-8.

Cogeneration permits connection of consumer owned and operated generating facilities to the utility's lines. Necessary protective devices must be installed and the unit coordinated and controlled by the utility. Cogeneration may present a hazard to workers performing work on the line they may believe de-energized. Multiple grounds should be placed on the line before work is begun.

PARALLELING THE SYSTEMS

Paralleling the consumer's generation facilities with those of the utility requires that additional protection equipment be installed at the

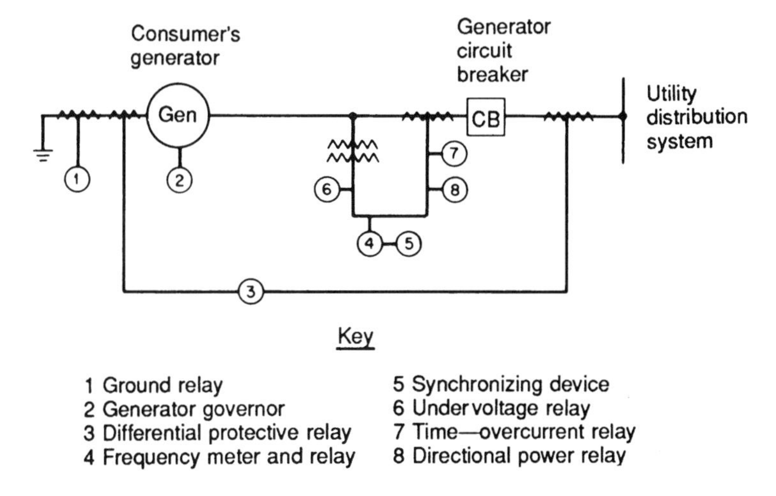

Figure 9-8. One-line diagram showing protection relaying for consumer cogeneration. *(Courtesy Long Island Lighting Co.)*

cogenerator's facilities. The principal features of this additional protection include:

1. Automatic synchronizing of the generator output with the utility,

2. Relaying to prevent the closing of the circuit breaker to the utility system until the cogenerator's generator is open, for protection of that generator,

3. Relaying to open the circuit breaker to the utility system on loss of power in the utility system,

4. Relaying to open the circuit breaker to the utility system on a ground fault on the utility system,

5. Relaying to control the cogenerator's generator circuit breaker to provide generator overcurrent protection, phase current balance protection, reverse power protection, under and over-frequency protection, and under and overvoltage protection,

6. Control of engine governor equipment for speed, generator phase match, and generator load.

The electrical connections and indicated protection are shown on the one-line diagram on Figure 9-8.

DISTRIBUTED GENERATION

A smaller version of cogeneration, known as distributed generation, was developed to be connected to the distribution systems at strategic points. These mainly consisted of small units, generally driven by small gas turbines, but may include wind, solar, fuel cell, and other experimental units; these may be both utility and consumer owned. These units, and some cogeneration units, are not usually competitive with utility owned larger units that have the advantage of scale.

Some cogeneration and distributed generation units may impact negatively on the safety of operations. Although standards for the selection, installation, and maintenance of equipment to connect and disconnect these units from the system to which they supply electric energy are furnished the consumer by the utility, these standards are not always followed, particularly those related to maintenance. This constitutes a hazard to persons who may be working on the systems; believing them to be deenergized, they may be the victims of an improper, unannounced, connection, energizing the systems to which they are connected. Similarly, should a fault develop on the utility system to which they are connected and the equipment fail to disconnect their generation from the system, overloads, fires and explosions may occur. Further, while they are under the supervision of the System Operator, they tend to dilute his attention to other events.

REMOTE METER READING AND DEMAND CONTROL

Electronic developments that have made e-mail (and the internet) inexpensive and universal means of communication have also made practical the remote reading of consumer's meters. Periodic inquiry automatically sent to each consumer identifies their meter and records the dial consumption and other data, transmitting it to the computer center where it may be automatically processed, producing the bill sent to the consumer.

In some cases, usually commercial and industrial consumers,

where it is desired to hold down their demands by arranging their loads not to coincide, and where practical to be scheduled for off peak hours (usually evening and early morning hours), the same means of communication is used to operate relays and switches to accomplish this purpose, often employing the same meter reading facilities.

REVIEW QUESTIONS

1. Of what parts does a distribution circuit consist?
2. What are the functions of the primary circuit? Of the secondary circuit?
3. What are some typical primary circuit voltages?
4. Name three kinds of distribution circuits.
5. What is the function of the opening in an open loop circuit?
6. What is a secondary network, and why?
7. What is the function of a network protector?
8. What is meant by cogeneration?
9. What special precautions should be taken in connecting cogeneration and distributed generation units to the distribution system?
10. What safety measures should be taken when cogeneration and distributed generation units are shut down?

Chapter 10

Essentials of Electricity

ELECTRICITY

When "electricity," is mentioned, it reflects broadly a form of energy that can perform many different kinds of work, the end result of which may be as varied as are the users of heat, light, mechanical, and chemical energy.

Electric current in a wire, on the other hand, identifies electricity in motion in a wire, just as the reference to "flow of water in a pipe" immediately identifies water in motion in a pipe (see Figure 10-1). Although comparison is made between electric current and water flow, the impression should not be had that electricity is a weightless fluid. Strange as it may seem, "electric current" in a wire is the simultaneous motion in the same direction of innumerable extremely small "particles of electricity" called *electrons.*

These electrons are already in the wire, but their normal motion is random, that is, as many are moving in one direction as in the other, so that there is no net movement in one direction, hence, there is no electric current.

In fact, when "electric pressure," which is referred to as "voltage," is applied, all that is done is to change the *random* motion of the electrons in the wire to a *directed* or *controlled* motion. (The correct meaning of voltage is *level* of electric pressure). By directed motion, what is meant is that all of the electrons move in the same direction.

Conductors and Insulators

The basis of all the concepts in the field of electricity is the *electron theory.* This states that all matter is composed of tiny units called *atoms.* These in turn subdivide into even smaller particles called *protons, electrons,* and *neutrons.* An atom is illustrated in Figure 10-2(a) as having a nucleus containing neutrons (having no electrical charge), and protons

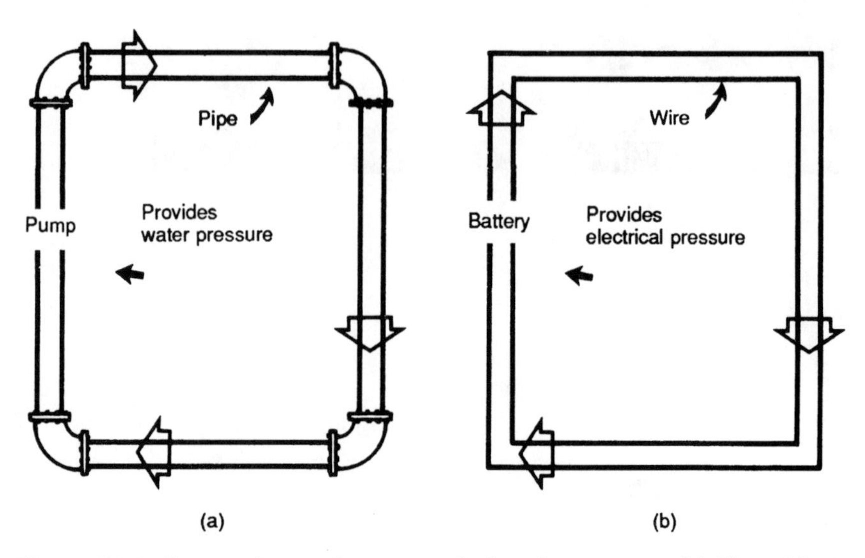

Figure 10-1. Comparison of water and electric currents. (a) Flow identifies water in motion. (b) Electric current identifies electricity in motion.

(having a *positive* electric charge). Around this nucleus spin electrons (having a *negative* electric charge) in various orbits. The electrons in the outer orbit of the atom may leave that orbit and move from atom to atom in a substance. These electrons are called *free electrons,* and normally move in all different directions. However, when a voltage is applied across a wire, the free electrons flow in one general direction, called the *current flow.* Materials having many free electrons are called *conductors,* and for a given amount of voltage will provide a large current flow.

Materials having few free electrons [see Figure 10-2(b)] are called *insulators* and for a given amount of voltage will provide little or no current flow. Typical good electrical conductors are copper and aluminum; good insulators are mica, porcelain, glass, and wood.

Resistance

Resistance is essentially the *opposition* offered by a substance to the flow of electric current. Remember that a conductor is made of a material whose atomic structure has a large number of free electrons. Because of the greater number of free electrons in their atomic structure, some conductors allow electric current to flow more readily than others. Re-

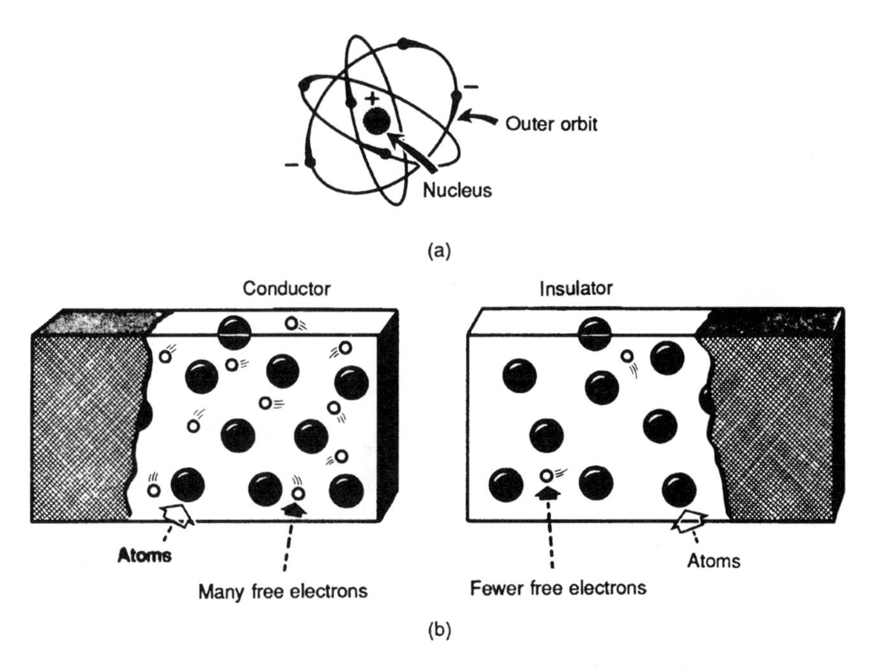

Figure 10-2. (a) The atom. An atom consists of a nucleus containing protons and neutrons around which planetary electrons revolve in orbits. (b) Comparison of conductor and insulator.

sistance in a conductor is similar to friction in a pipe. A pipe with a smooth inside surface conducts water with little loss of pressure or hindrance to the flow. But if the pipe is rough inside, the pressure and flow are reduced. Similarly, a good conductor allows electricity to flow with little opposition or resistance and with a small loss of electrical pressure or voltage. A poor conductor offers a large resistance to electric current and this causes a large loss of pressure or voltage. Just as the length and cross section of the pipe affect the flow of water so it is with conductors of electric current: the greater the length, the greater the resistance; the greater the cross-sectional area, the smaller the resistance (see Figure 10-3). Additionally the material of which the conductor is made also affects the resistance. The energy used in overcoming resistance is converted into heat that is why some conductors are found to be warm when touched.

The Electric Circuit

Electric current flows only when wires or metallic conductors form

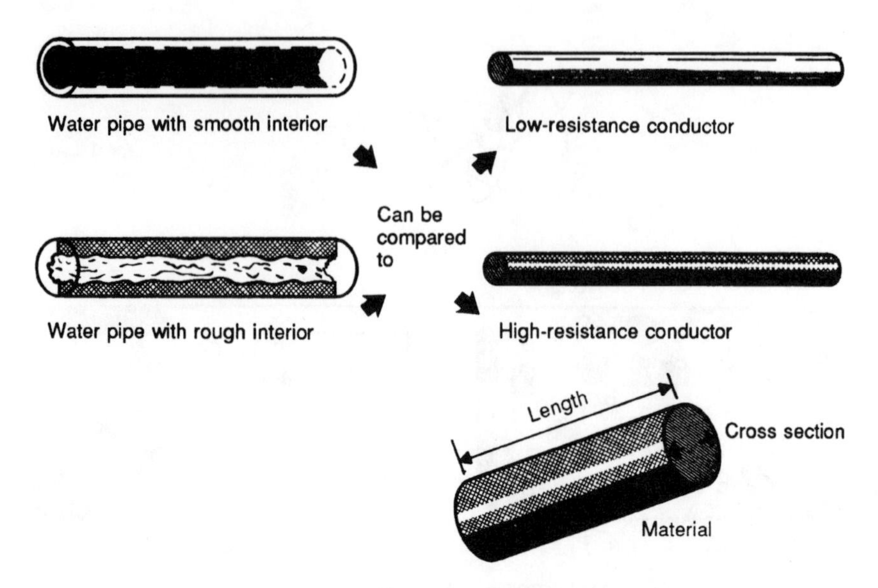

Figure 10-3. Resistance.

a *closed path.* These closed paths can be arranged in several ways—in series, parallel, and series-parallel—and are referred to as circuits, or *electric circuits.* A series circuit [see Figure 10-4(a)] is formed by connecting a storage battery to a number of lamps connected as shown. If for any reason one of the lamps goes out, there is no longer a closed electric circuit and current will not flow.

If the same storage battery is taken and the lamps are connected across each other, a parallel circuit is formed, shown in Figure 10-4(b). In this arrangement should any one lamp go out, the remainder will remain lit. By combining lamps in series with those in parallel the series-parallel circuit is formed, shown in Figure 10-4(c). In this circuit, the series section has all the characteristics of the series circuit, and the parallel section has the characteristics of the parallel circuit.

Difference of Pressure— Difference of Potential-Voltage

In a water system that is made up of a water pump, water pipes, and a valve (to turn the water on and off), water will flow only when the pump is working. A pump enables a flow of water by virtue of the fact that it creates a difference in pressure, causing the water to flow from a point in the pipe at high pressure to a point in the pipe at a lower pres-

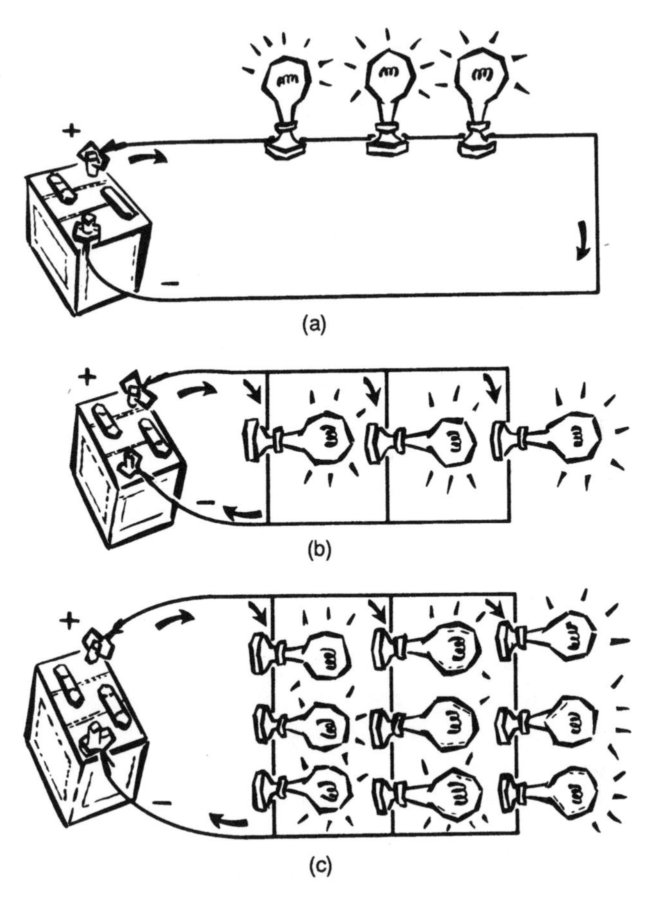

Figure 10-4. Lamps maybe connected to form, (a) a series circuit, (b) a parallel circuit, and (c) a series-parallel circuit.

sure. Thus, with the pump rotating and with the valve of the water pipe turned on, a steady flow of water will be obtained. If the pump rotates and if the valve is in the "off" position, a relatively large amount of pressure is built up near the valve; but no water will flow.

The water system just discussed can be compared to an electric system (see Figure 10-5), replacing the pump with a generator, the pipe with a conductor, and the valve with a switch to turn the electric current on and off. By analogy, to obtain current flow in the electrical system, the generator must be rotating and the switch must be placed in the on position. Rotating the generator creates a "difference of electric pressure," also called difference of potential, at the terminals of the generator. This is the difference of electric pressure which will cause the electrons in the conductor to move, that is, cause the current to flow in a conductor. If the switch is placed in the off position, while the generator is rotating, electric pressure or potential will be built up near the switch terminals but no electric current will flow.

Electrical Quantities

Resistance, the opposition offered by a substance to the flow of current; *current,* the flow of electrons through a conductor; and *voltage,* the electrical pressure that forces current to flow through a resistance, all have been discussed. These can now be discussed further regarding the quantities or units in which each are measured. Resistance is measured in *ohms,* current in *amperes,* and voltage in volts. These may be tied together with the basic statement that one ampere of current will flow through a resistance of one ohm when a voltage of one volt is applied across the resistance [see Figure 10-6(a)].

The unit of electrical power is the watt, with 746 watts considered equal to one horsepower. The watt is equal to the current in a dc circuit multiplied by the voltage applied to the circuit. Thus, if 100 amperes flow in a circuit in which there is a 10-volt battery, the battery is supplying 1000 watts of power. Another way of saying this is that the load is consuming 1000 watts.

Electrical power consumed in the home or industry is measured in

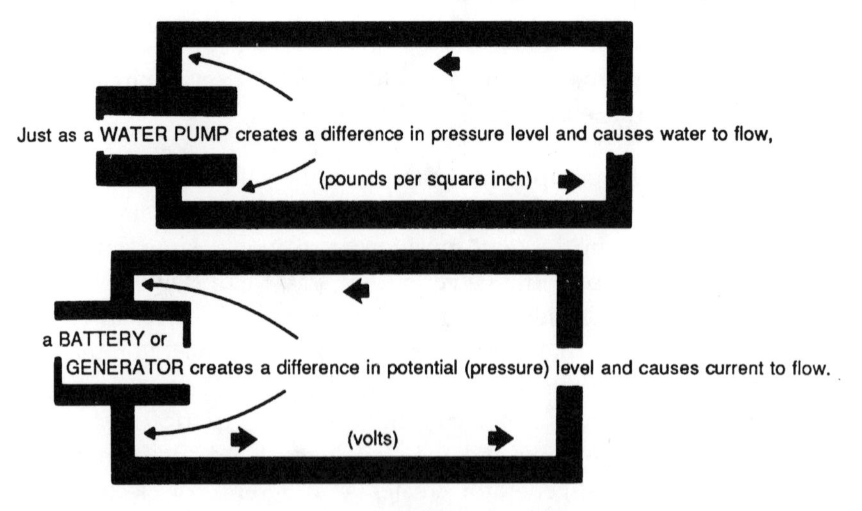

Figure 10-5. Voltage is a measure of difference of potential.

terms of watt-hours. This is equal to the power consumed by a circuit multiplied by the number of hours in which it was consumed. The instrument that measures this is called a *watt-hour meter*, and electric bills are calculated on the basis of readings made on this meter [Figure 10-6(b)].

While the basic units used in the measurement of electrical quantities are the ohm, volt, ampere, and watt, these units often appear in such large quantities that it would be inconvenient to speak in terms of the basic unit. For example, approximately 6 volts may be obtained from a storage battery, and 117 volts from a house electrical receptacle. However, when discussing high-voltage transformers and transmission lines terms of 10,000 V, 50,000 V, and sometimes voltage in excess of 100,000 V are used. To simplify this, the prefix "kilo" is generally used for large quantities. This indicates that the number should be multiplied by 1000. Thus, a 10-kilovolt (kV) transformer is rated as being capable of handling 10,000 V; similarly, 60 kV, 60 kW, and so on. In general notation, abbreviate kilo and simply refer to "k," hence, kV, kW, and so on.

Another commonly used prefix is the term "mega" or "meg" which indicates that the number associated with it must be multiplied by 1,000,000. In general use this term is too large to be associated with either current or voltage, but is often used in connection with insulators in such expressions as 1 megohm or 50 megohms.

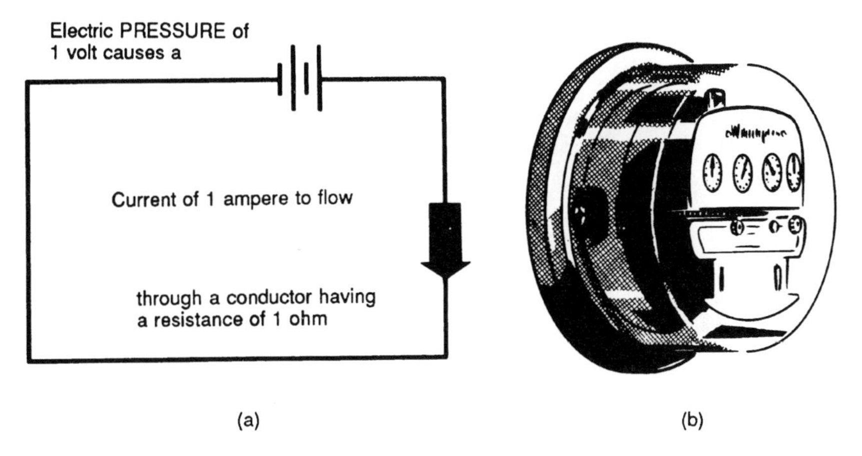

Figure 10-6. Electrical quantities. (a) The power in this circuit is: 1 volt × 1 ampere 1 watt. (b) In 1000 hours, a watt-hour meter would read 1 kilowatt hour.

Two other prefixes sometimes used are "mil" and "milli" for one thousandth of some quantity, and "micro" for one millionth of some quantity. Examples of these are millivolt for 0.001 V and microvolt (μV) for 0. 000001 V. Actually, these various prefixes represent nothing more than a "shorthand" notation; they make the task of discussing very large or very small quantities easier. Figure 10-7 illustrates the shorthand notations.

Ohm's Law

The relationship of current, voltage, and resistance must be considered in every electric circuit. Direct current will flow only in a closed circuit, one which provides a continuous conducting path from the negative to the positive terminal of the voltage source. The way the electromotive force (emf), E, is distributed through the circuit and the relationships of voltage, current, and resistance, are contained in Ohm's law, which states that *the current in an electric circuit is directly proportional to the voltage and inversely proportional to the resistance.*

This law may be expressed by the equation

1. $I = E/R$ or current equals the voltage divided by the resistance.

From which equations 2 and 3 may be derived:

2. $R = E/I$, or resistance equals the voltage divided by the current.

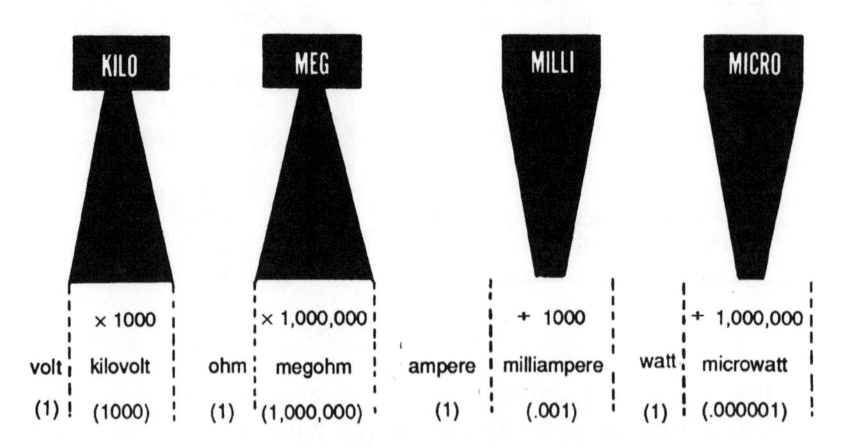

Figure 10-7. Shorthand notation.

3. E = IR, or the voltage is equal to the product of the current times the resistance.

For example: with a battery voltage of 12 volts (shown on the voltmeter) and a resistance of 3 ohms, the amperes may be found as follows:

$$I = E/R$$
$$I = 12 \text{ volts} \div 3 \text{ ohms}$$
$$I = 4 \text{ amperes or } 4 \text{ A}$$

Figure 10-8 illustrates pictorially the relationships between E, I, and R.

Alternating Current

So far, we have discussed the transfer of energy from one point to another point, both through the medium of water and electricity circulating continuously in a circuit in *one* direction.

Now, consider what happens when the pump in Figure 10-9 is actuated from an outside source of power. Water in pipe *B* will be pushed in one direction causing a flow of water or current in that direction. Assuming the valve is open, the water will push the piston in the water motor in the same direction. Now, as the piston of the pump moves in the opposite direction, water in pipe *A* will be pushed, causing a flow of water in the direction of the water motor. The water will push the piston of the water motor in the direction opposite to that previously

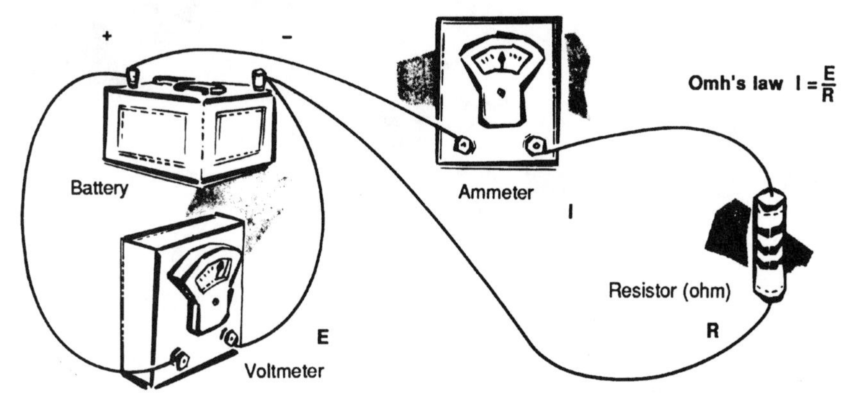

Figure 10-8. Pictorial representation of E, I, R, and Ohm's Law.

moved. Thus, an oscillating, or to-and-fro, motion is transmitted from one end of the circuit to the other. The piston water motor can then be used to drive machinery.

It is seen, therefore, that complete circulation of water or electricity is not necessary to transmit energy from one point to another. In this method, currents alternate their direction periodically, first flowing in one direction in a pipe, and then in the other. In electric circuits, this is called an *alternating current* (ac) system to distinguish it from the continuous or direct current (dc) system. Utility companies almost always use alternating current for their distribution systems.

Power in an AC Circuit

To obtain the power in a dc circuit, the voltage in volts is simply multiplied by the current in amperes. Thus, the power in watts is obtained. The electrical energy in the dc circuit is transformed into heat at the rate of $P = 1^2R$, where P is power, I is current in amperes, and R is resistance in ohms.

The power consumed in the ac circuit is the average of all the values of instantaneous power or the rate at which electrical energy is transformed into heat or mechanical energy for a complete cycle. To find the power value, the instantaneous voltage and current value are mul-

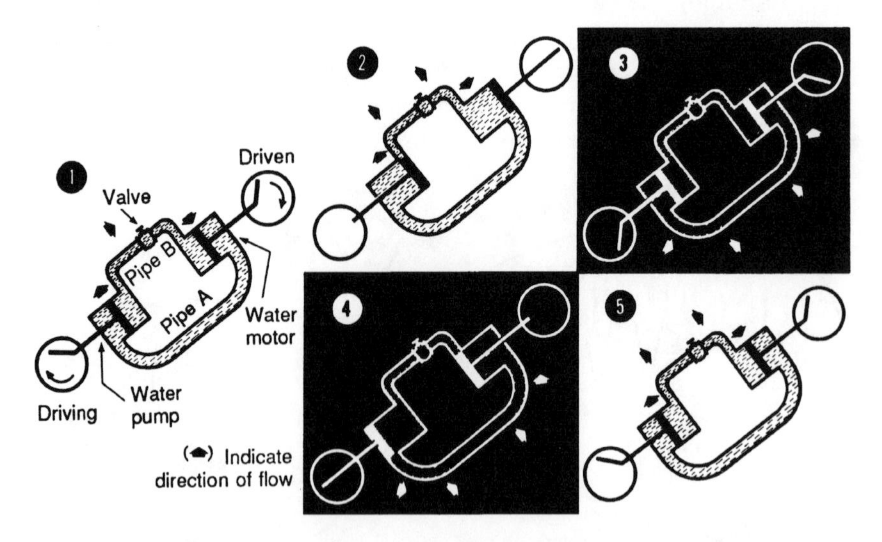

Figure 10-9. Water analogy of alternating current. A pumping system in which the energy of motion is transferred by water.

tiplied and the power value for each instant is plotted on a graph. The power curve obtained shows the value of instantaneous power as a function of time. The average of the instantaneous power obtained from the curve is the average power absorbed in the circuit. In an ac circuit which contains resistance only, the average power is equal to I, the maximum instantaneous power. The average power value of such a circuit is equal to the product E effective × I effective. (E is voltage and I is current.)

The effective value of ac is that value that produces the same amount of work as an equivalent amount of dc. When the ac circuit contains resistance only, the product E effective × I effective (the power in watts) is the same as in the corresponding dc circuit. In ac circuits which contain inductance and capacitance, the product E effective × I effective is not equal to the power in watts but is called the *apparent* power expressed in volt-amperes. The power in watts for these circuits is equal to 1^2R or E^2/R. This is also called the *true* power of the circuit, and it is absorbed in the resistive part of the circuit.

The ratio between the *true* power in watts and the *apparent* power in volt-amperes is called the *power factor* of the circuit. For the theoretical circuit shown in Figure 10-10, containing resistance only, the power factor is equal to 1. Practically, the power factor is always less than 1 because all circuits contain inductance and capacitance in some degree. It is in the interest of the utility company and of the consumer to maintain a power factor as close to 1 as possible.

Electromagnetic Induction—Alternating Current

If a conductor which is part of a closed circuit is moved through a

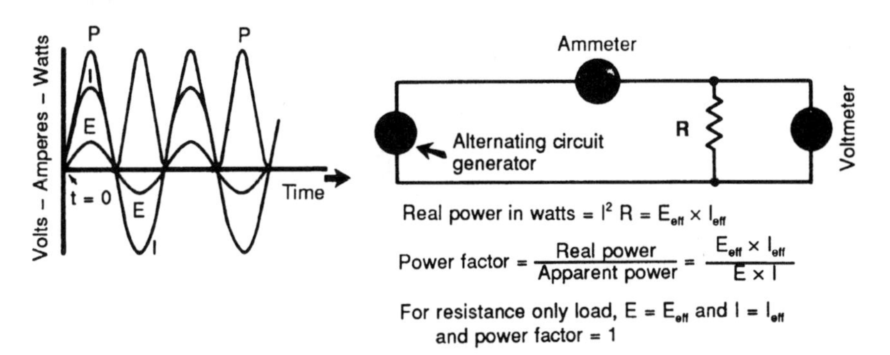

Figure 10-10. Power in an ac circuit. Circuit containing resistance only.

magnetic field, an electric current will flow in the conductor. The current results from what is called the *induced* electromotive force (emf). But this happens only if the directions (1) of the current, (2) of the magnetic field, and (3) of the motion are at right angles to each other.

If a loop of wire as illustrated in Figure 10-11 is rotated at uniform speed in a magnetic field, a voltage is induced in the conductor that makes up the loop. By measuring the voltage for different positions of the loops and recording the angular position of the loop, the waveform of the induced voltage is obtained. This waveform is called a *sine wave.* Note that the instantaneous values of the voltage change continuously from zero to maximum, from maximum back to zero, from zero to a minimum, and from the minimum back to zero, thus completing one cycle. The maximum, or the minimum of the voltage wave, is called the *amplitude* of the voltage, and the number of cycles per second the *frequency of* the wave. The part of the curve from zero-to-maximum back to zero is called an *alternation.* Such a waveform is an example of the form of voltage and current in the alternating current circuit.

Magnetic Field Around a Conductor

When an electric current flows through a conductor, a magnetic

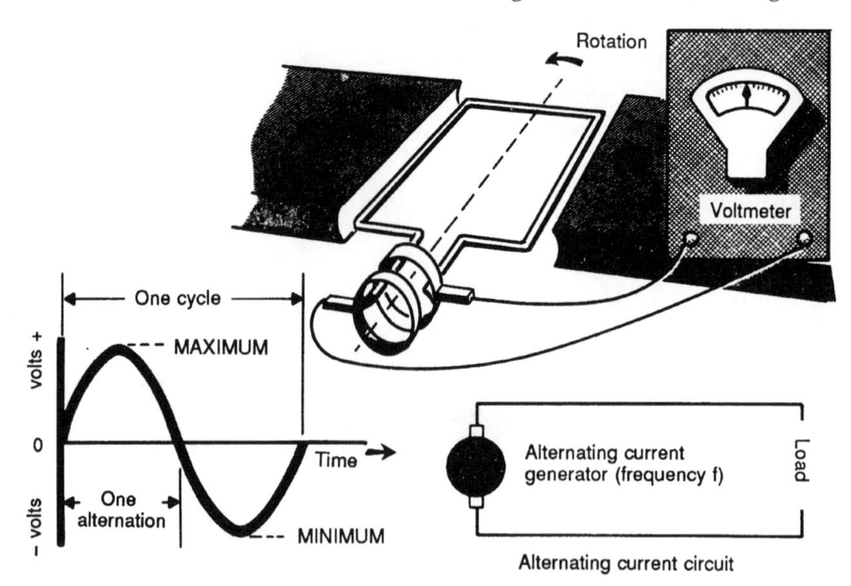

Figure 10-11. Sine-wave voltage is induced in coil as it rotates in magnetic field.

field is created. The lines of force of this magnetic field form concentric circles around the conductor (see Figure 10-12) with the strength of this field depending on the amount of current flowing in the conductor. If the conductor is wound in a coil formed of many turns as shown in Figure 10-13, there will be an increase in the number of the lines of force because now each turn of the coil aids every other turn in strengthening the magnetic field.

An increase in the lines of force, or strength of the magnetic field, may be further obtained by winding the coil around a closed iron core (see Figure 10-14). This occurs because the iron core provides an easy path (compared to air) for the lines of force of the magnetic field.

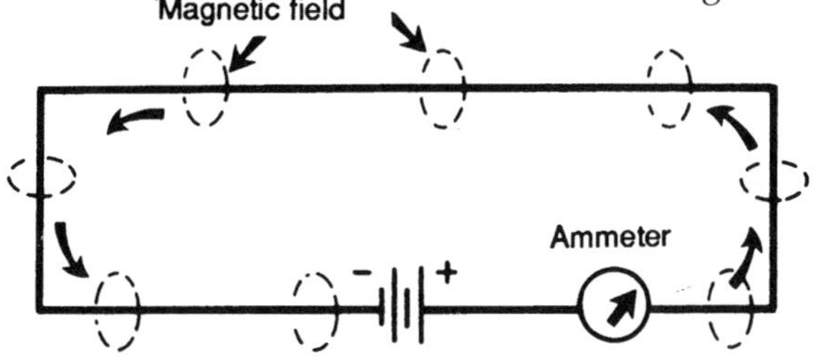

Figure 10-12. Current carrying conductor surrounded by a magnetic field.

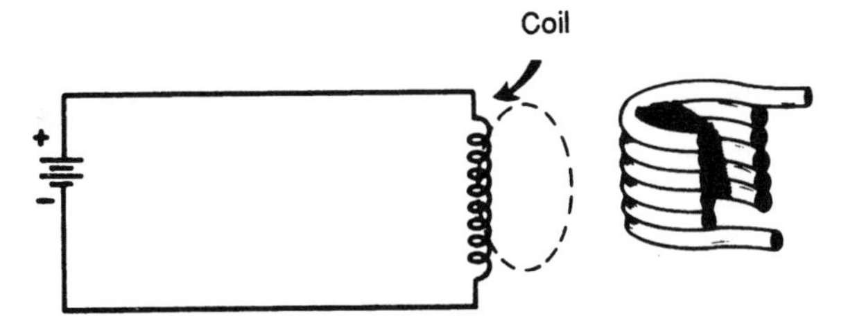

Figure 10-13. Magnetic field is strengthened by forming a coil.

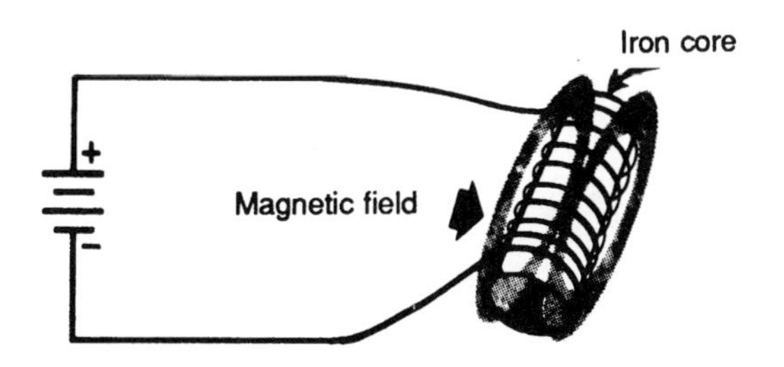

Figure 10-14. Magnetic field is further strengthened by winding coil around an iron core.

Polyphase Voltage

The generation of a sine wave just described was the result of rotating a coil of wire in the magnetic field set up by two poles of a magnet. In many cases it is more efficient and more economical to generate a voltage by using many coils, each electrically insulated from each other. When this is done, each coil produces a sine wave and the output of such a generator appears as a series of sine waves each slightly displaced. A sine wave may be divided into 360 degrees, and called a complete cycle. Using this idea, if three coils are rotated in the magnetic field, and if they are separated by 120 degrees, the sine-wave output will be a series of sine waves each separated by 120 degrees [see Figure 10-15(a) and 10-15(b)]. This is called *three-phase voltage* as against single-phase voltage when a single coil is used.

In practice, the use of single or three-phase voltage is a matter of

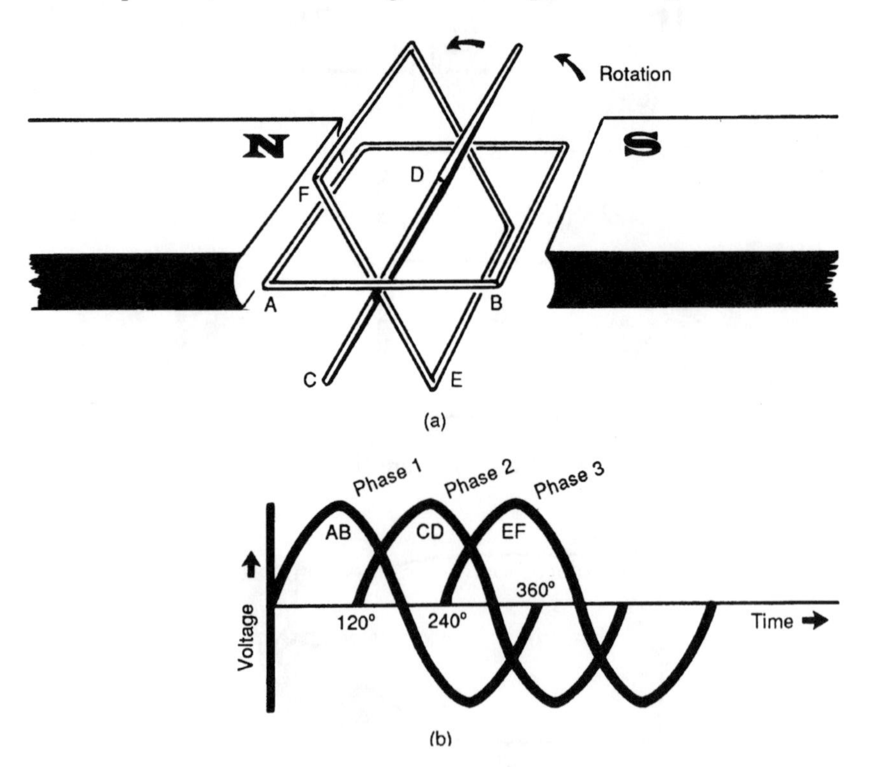

Figure 10-15. Polyphase voltage. (a) Simplified diagram of 3-phase generator and (b) output waveform of 3-phase generator.

considerable importance. However, in the basic discussion of the equipment and concepts involved in power distribution, the phase of the generated electricity is of no great importance, except on a few occasions where the reader will find it mentioned.

Induced Voltages

If another coil is wound on the same iron core, the lines of force of the magnetic field created by the current flowing through the first coil cut through the turns of the second coil. The closer the two coils are, the greater will be the number of lines of force that intercept the turns of the second coil. If a voltmeter is connected across the second coil the meter will show a deflection every time the switch is closed or opened. The voltage that is measured in this second coil is called an *induced voltage* (see Figure 10-16).

From this is derived a very important law of electricity: *Whenever magnetic lines of force cut through a conductor, a voltage is generated or induced in that conductor.* Since there is no *change* in the current flow while the switch is closed or open, no voltage is induced into the second coil during this time. Suppose the dc battery is replaced with a source of alternating current. This will produce a current flow that is continually changing in value, which in turn produces a continuously changing magnetic field. The voltmeter will now read a continuous alternating current. There are now the ingredients of a device called a *transformer.* The coil to which the ac voltage is applied is called the *primary,* with the second coil on the core called the *secondary.*

HOW THE TRANSFORMER WORKS

The Transformer

Fundamentally, as shown in Figure 10-17, a transformer consists of two or more windings placed on a common iron core. All transformers have a primary winding and one or more secondary windings. The core of a transformer is made of laminated iron and links the coils of insulated wire that are wound around it. There is no electrical connection between the primary and the secondary; the *coupling* between them is through magnetic fields. This is why transformers are sometimes used for no other purpose than to isolate one circuit from another electrically. When this is done, the transformer used for this purpose is called an

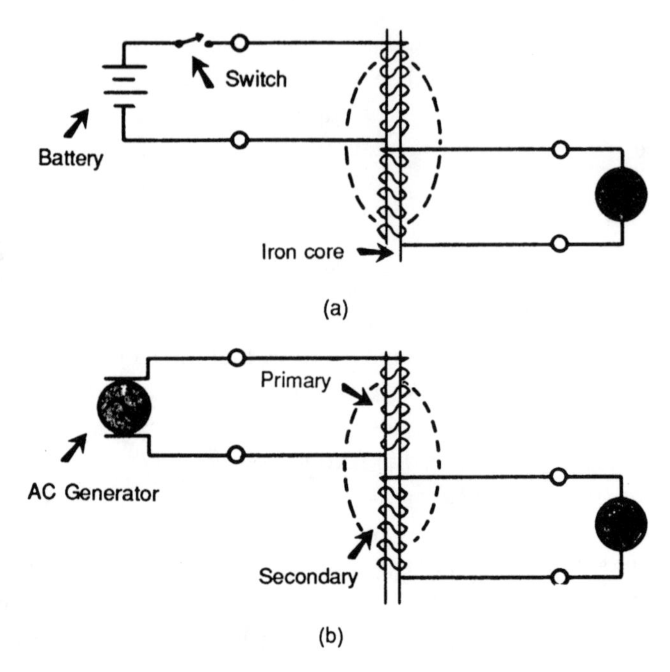

Figure 10-16. Induced voltage. (a) When switch is opened or closed, the voltmeter will deflect. (b) Voltmeter reads continuous ac voltage.

iso*lating* transformer.

The winding that is connected to the source of power is called the *primary,* and the winding connected to the *load* is called the *secondary.* The essential function of the conventional power transformer is to transfer power from the primary to the secondary with a minimum of losses. In the process of transferring energy from primary to secondary, the voltage delivered to the load may be made higher or lower than the primary voltage. The load may be some household appliance, an industrial electrical device, or even another transformer.

Step-Up and Step-Down Transformers

Transformers are wound to be as close to 100 percent efficient as possible. That is, all the power in the primary should be transferred to the secondary. This is done by selecting the proper core material, winding the primary and secondary close to each other, and a number of other careful designs.

Thus, assuming 100 percent transformer efficiency, it can then be

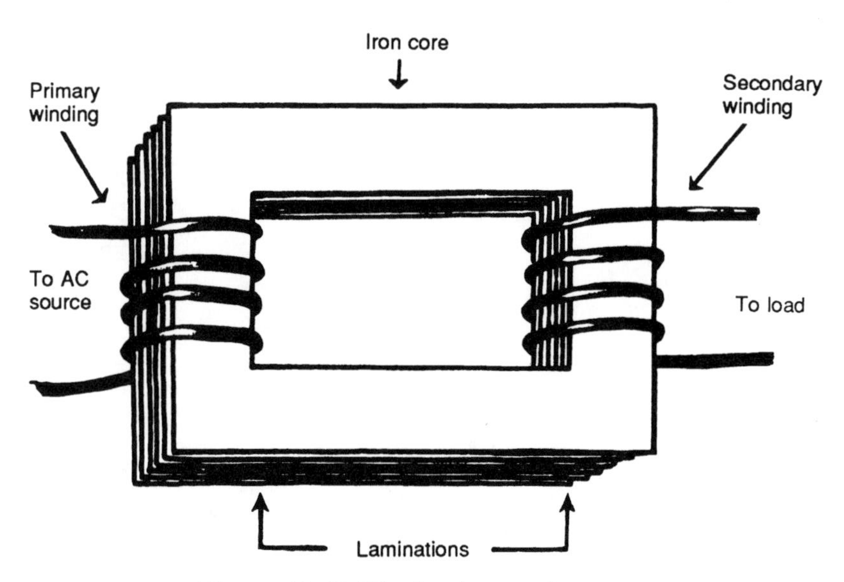

Figure 10-17. The basic transformer.

assumed that the relationship between the primary and secondary voltages will be the same as the relationship between their turns. If the secondary has *more* turns than the primary it can be said that the transformer is operating as a *step-up* transformer; if the secondary *has fewer turns* than the primary we say that the transformer is operating as a *step-down* transformer.

Figure 10-18 gives a representation of an autotransformer together with a conventional transformer. The autotransformer is somewhat different from the conventional transformer in that portions of the primary and secondary are *common,* or make use of the same turns. However, like all other transformers, the autotransformer does have the basic primary and secondary windings.

Turns Ratio

Whether a transformer is of a step-up or step-down type the power in the primary is equal to the power in the secondary. Thus, if the load draws 1000 watts, the product of voltage and current in the primary is also equal to 1000 watts. Another important principle is the fact that the primary and secondary voltages are in the same ratio as their turns. If the secondary has twice the turns of the primary, the secondary voltage will be twice as great as that of the primary. Figure 10-19 illustrates the significance of turns ratio.

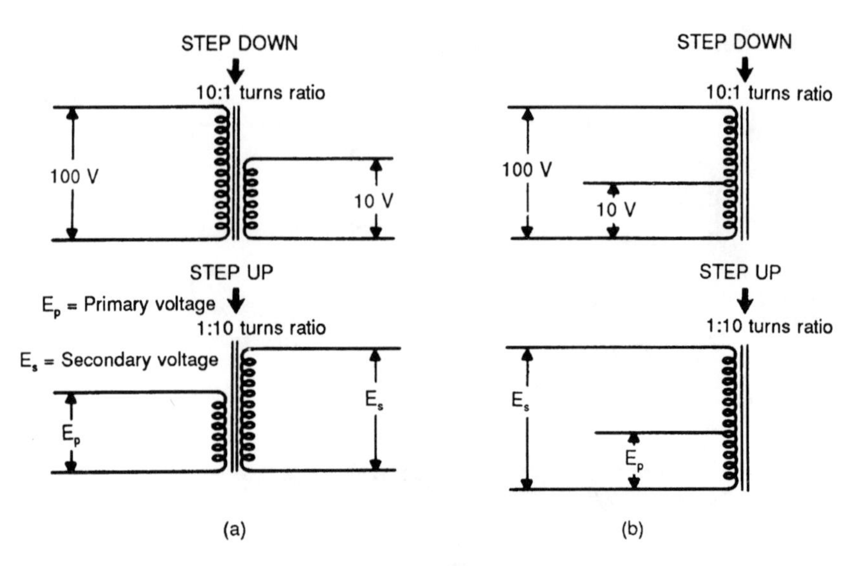

Figure 10-18. Comparison of (a) a conventional transformer and (b) an autotransformer.

Transformer Rating

The nameplate on a transformer (see Figure 10-20) gives all the pertinent information required for the proper operation and maintenance of the unit. The capacity of a transformer (or any other piece of electrical equipment) is limited by the permissible temperature rise during operation. The heat generated in a transformer is determined by both the current and the voltage. Of more importance is the kilovolt-ampere rating of the transformer. This indicates the maximum power on which the transformer is designed to operate under normal conditions (when the current and voltage are "in phase"). Other information generally given is the phase (single-phase, three-phase, etc.), the primary and secondary voltages, frequency, the permitted temperature rise, and the cooling requirements—which include the number of gallons of fluid that the cooling tanks may hold. Primary and secondary currents may be stated at full load.

Depending on the type of transformer and its special applications, there may be other types of identifications for various gages, temperature indication, pressure, drains, and various valves.

Thus, it can be seen that while the transformer consists primarily

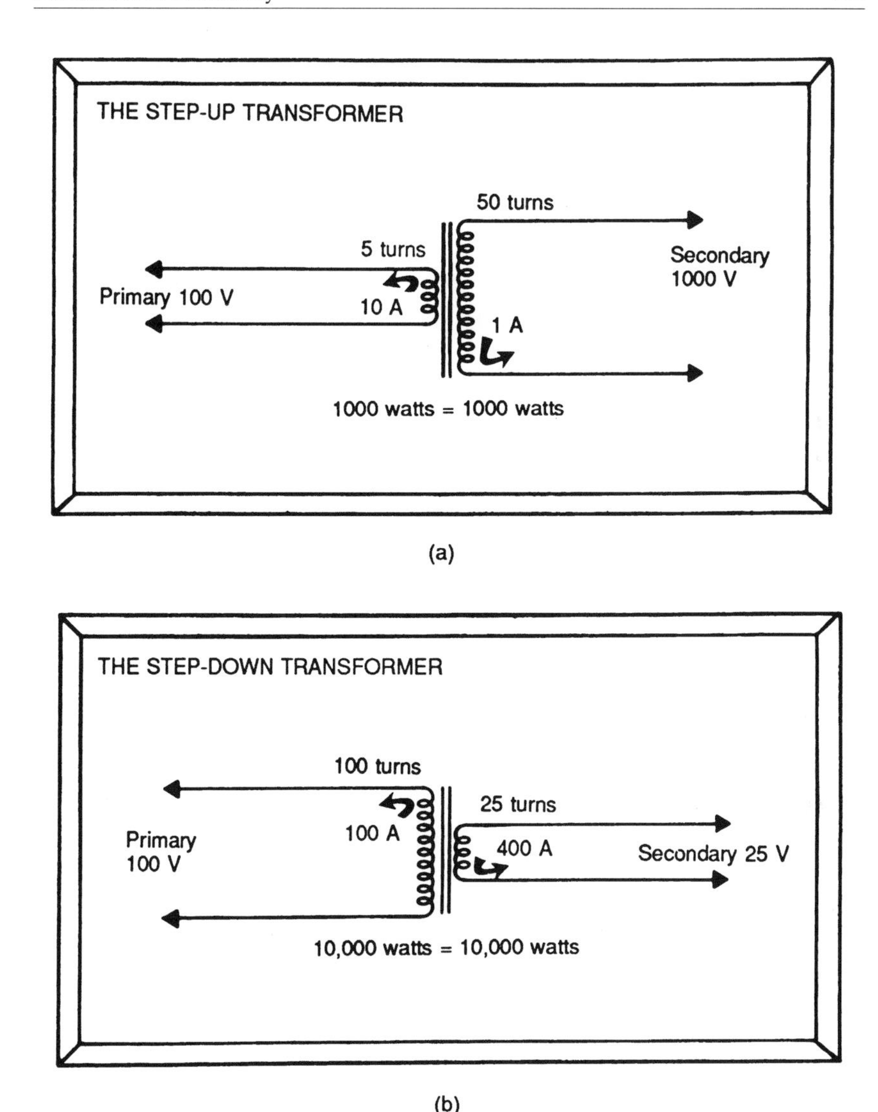

Figure 10-19. (a) The step-up and (b) step-down transformers.

of a primary and a secondary winding, there are many other points to take into consideration when selecting a transformer for a particular use.

Methods of Transformer Cooling

The wasted energy in the form of heat generated in transformers due to unpreventable iron and copper losses must be carried away to

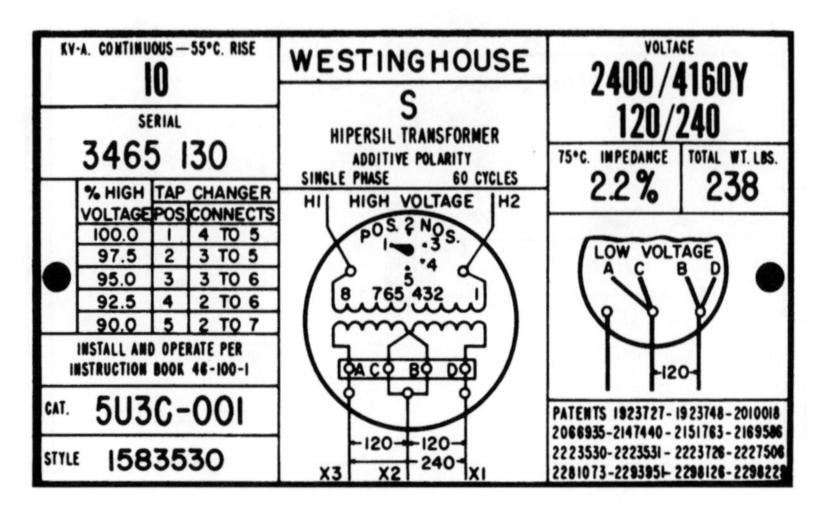

Figure 10-20. Transformer nameplate.

prevent excessive rise of temperature and injury to the insulation about the conductors. The cooling method used must be capable of maintaining a sufficiently low average temperature. It must also be capable of preventing an excessive temperature rise in any portion of the transformer, and the formation of "hot spots." This is accomplished, for example, by submerging the core and coils of the transformer in oil, and allowing free circulation for the oil [Figure 10-21(a)]. Sometimes, for reasons of safety, the use of oil as a cooling agent is prohibited. (Oil can be a fire hazard.) In these situations, special fluids, known as "askarels" can be used in place of oil. However, when using these substitutes, the varnishes which are generally applied to the insulation of the coils must be chosen carefully. For indoor use, in clean, dry locations, open dry-type air cooling can be used [see Figure 10-21 (b)]. For outdoor or indoor, use, a sealed dry type can be obtained.

Some transformers are cooled by other means: (a) by forced air or air blast, (b) by a combination of forced oil and forced air, and (c) in some special applications, by water cooling.

REVIEW QUESTIONS

1. Define electric current.
2. Name two materials that make good typical electrical conductors,

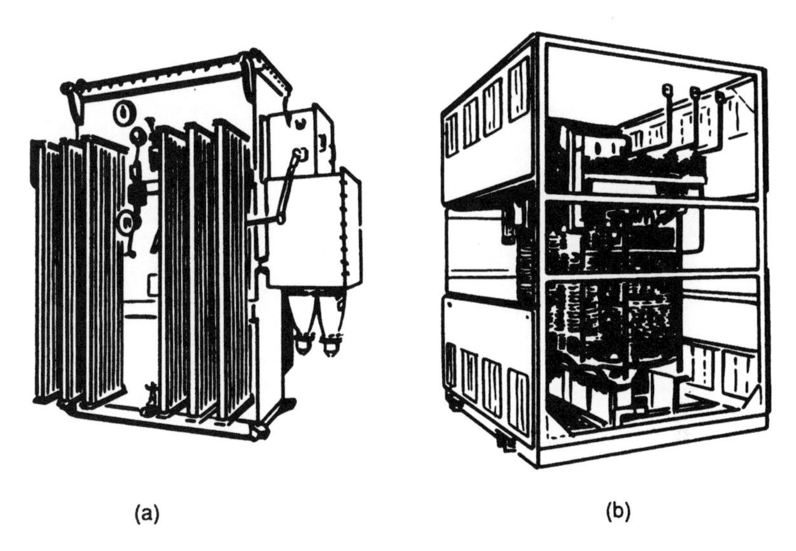

Figure 10-21. Methods of cooling transformers. (a) Oil-filled and (b) air-cooled dry type.

and two materials that make good electrical insulators.

3. What is meant by resistance to electrical current?
4. Name three types of electric circuits.
5. What is meant by the term "difference of potential"?
6. What are the units used to measure electrical voltage, current, resistance, and power.
7. Explain the use of the prefixes: micro, kilo, mega; use the volt as a base unit.
8. State the three forms of Ohm's law.
9. What is the power factor of an alternating-current circuit?
10. Explain the turns ratio of a transformer.

Appendix A

Insulation: Porcelain Vs. Polymer

For many years, porcelain for insulation purposes on lines and equipment has exercised a virtual monopoly. It was perhaps inevitable that plastics, successful as insulation for conductors since the early 1950's, should supplant porcelain as insulation for other applications in the electric power field.

The positive properties of porcelain are chiefly its high insulation value and its great strength under compression. Its negative features are its weight (low strength to weight ratio) and its tendency to fragmentation under stress. Much of the strength of a porcelain insulator is consumed in supporting its own weight. Figure A-1a&b.

In contrast, the so-called polymer not only has equally high insulation value, but acceptable strength under both compression and tension. It also has better water and sleet shedding properties, hence handles contamination more effectively, and is less prone to damage or destruction from vandalism. It is very much lighter in weight than porcelain (better strength to weight ratio), therefore more easily handled. Figure A-2, Table A-1.

Economically, costs of porcelain and polymer materials are very competitive, but the handling factors very much favor the polymer.

Polymer insulation is generally associated with a mechanically stronger insulation, such as high strength fiberglass. The fiberglass insulation serves as an internal structure around which the polymer insulation is attached, usually in the form (and function) of petticoats (sometimes also referred to as bands, water shedders; but for comparison purposes, however, here only the term petticoat will be used). The insulation value of the Polymer petticoats is equal to or greater than that of the fiberglass to which it is attached.

The internal fiberglass structure may take the form of a rod (or shaft), a tube, cylinder, or other shape. It has a high comparable com-

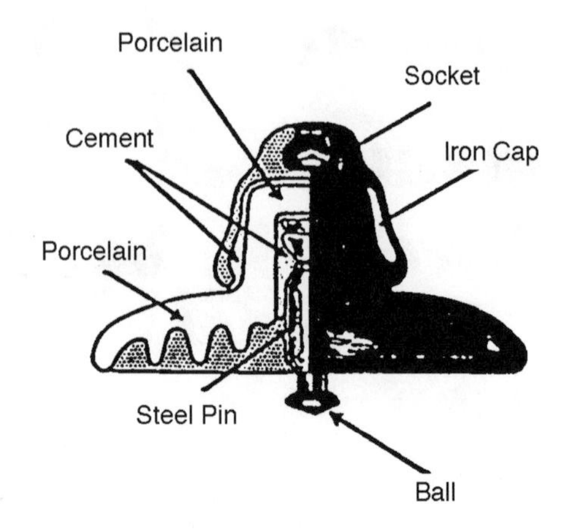

Figure A-1 a. Ball and socket type suspension insulator.

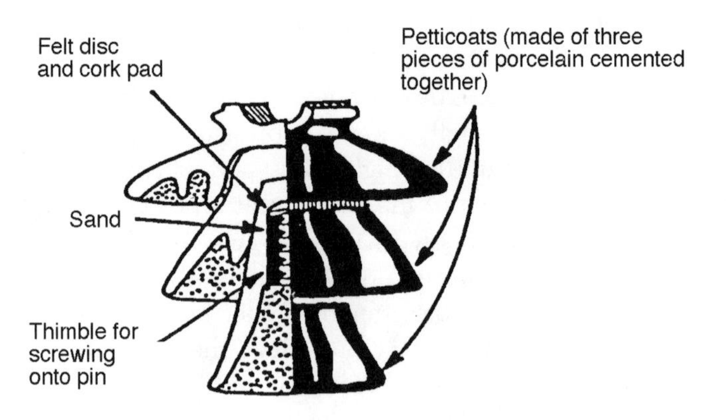

Figure A-1 b. Pin-type insulator.

pression strength as a solid and its tensile strength, equally high, is further improved by stranding and aligning around a fiber center. The polymer petticoats are installed around the fiberglass insulation and sealed to prevent moisture or contamination from entering between the petticoats and fiberglass; Figure A-3. The metal fittings at either end are crimped directly to the fiberglass, developing a high percentage of the inherent strength of the fiberglass. It should be noted that fiberglass with an elastometric (plastic) covering has been used for insulation purposes since the early 1920's.

The polymer petticoats serve the same function as the petticoats associated with porcelain insulators, that of providing a greater path for electric leakage between the energized conductors (terminals, buses,

Table A-1. Polymer insulation weight advantage

Product	*Type*	*Voltage (kV)*	*Porcelain Weight (lbs)*	*Polymer Weight (lbs)*	*Percent Weight Reduction*
Insulator	Distribution	15	9.5	2.4	74.7
Arrester	Distribution	15	6.0	3.8	36.7
Post Insulator	Transmission	69	82.5	27.2	67.0
Suspension	Transmission	138	119.0	8.0	93.2
Intermediate Arrester	Substation	69	124.0	28.0	77.4
Station Arrester	Substation	138	280.0	98.9	64.7

Figure A-2. A variety of typical Polymer insulator shapes. (*Courtesy Hubbell Power Systems*)

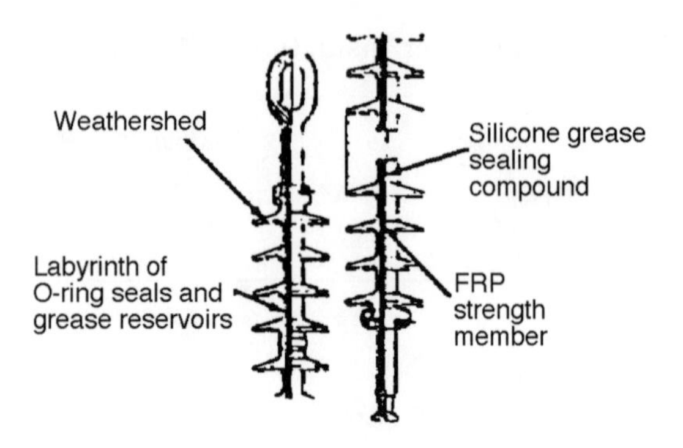

Figure A-3. Polymer insulator showing fiberglass rod insulation and sealing (Courtesy Hubbell Power Systems).

etc.) and the supporting structures. In inclement weather, this involves the shedding of rain water or sleet as readily as possible to maintain as much as possible the electric resistance between the energized element and the supporting structure, so that the leakage of electrical current between these two points be kept as low as possible to prevent flashover and possible damage or destruction of the insulator. Tables A-2a&b.

When the insulator becomes wet, and especially in a contaminated environment, leakage currents begin to flow on the surface; if the current becomes high enough, an external flashover takes place. The rate at which the insulation dries is critical. The relationship between the outer petticoat diameter and the core is known as the form factor. The leakage current generates heat (I^2R) on the surface of the insulator (eddy currents). In addition to the effects of the leakage current, the rate at which the petticoat insulation will dry depends on a number of factors. Starting with its contamination before becoming wet, the temperature and humidity of the atmosphere and wind velocity following the cessation of the inclement weather. In areas where extreme contamination may occur (such as some industrial areas or proximity to ocean salt spray), the polymer petticoats may be alternated in different sizes, Figure A-4, to obtain greater distance between the outer edges of the petticoats across which flashover might occur. When dry, the leakage current (approximately) ceases and the line voltage is supported across dry petticoats, preventing flashover of the insulator. It is obviously impractical to design and manufacture comparable porcelain insulators as thin as polymers and having the same form factor. Table A-3.

In addition, in porcelain insulators, the active insulating segment is usually small and, when subjected to lightning or surge voltage stresses, may be punctured. In subsequent similar circumstances, it may breakdown completely, not only causing flashover between the energized element and the supporting structure, but may explode causing porcelain fragmentation in the process; the one-piece fiberglass insulator will not experience puncture.

The polymer suitable for high voltage application consists of these materials:

1. Ethylene Propylene Monomer (EPM)
2. Ethylene Propylene Diene Monomer (EPDM)
3. Silicone Rubber (SR)

Both EPM and EPDM, jointly referred to as EP, are known for their inherent resistance to tracking and corrosion, and for their physical properties, SR offers good contamination performance and resistance to Ultra Violet (UV) sun rays. The result of combining these is a product that achieves the water repellent feature (hydrophobic) of silicone and the electromechanical advantages of EP rubber.

Table A-2a. Polymer improvement over porcelain (*Courtesy Hubbell Power Systems*)

Watts Loss Reduction* (watts per insulator string)					
Voltage			*Relative Humidity*		
kV	*30%*	*50%*	*70%*	*90%*	*100%*
69	0.6-	0.8	0.9	1.0	4.0
138	1.0	2.4	4.5	7.2	8.0
230	1.0	2.5	5.7	14.0	29.0
345	2.5	4.2	8.5	15.0	30.0
500	2.8	7.8	11.5	33.0	56.0

*Power loss measurements under dynamic humidity conditions on I-strings.

Table A-2b. Polymer Distribution Arrester Leakage Distance Advantage ***(Courtesy Hubbell Power Systems)***

Standard MCOV	*Standard Porcelain Leakage Distance (in)*	*Special Porcelain Height (inches)*	*Special Porcelain Leakage Distance (in)*	*Standard Porcelain Height (inches)*	*Standard Polymer Leakage Distance (in)*	*Polymer Height (inches)*
8.4	9.0	9.4	18.3	15.9	15.4	5.5
15.3	18.3	15.9	22.0	20.0	26.0	8.5
22.0	22.0	20.0	29.0	28.9	52.0	17.2

Table A-3. Comparison of Contamination Performance of Polymer versus Porcelain Housed Intermediate Class Arresters ***(Courtesy Hubbell Power Systems)***

			Partial Wetting Test		*5 Hr. Slurry Test Maximum Current After Slurry Number*			
MCOV (kV)	*Housing Material*	*Housing Leakage Distance (in)*	*Max. Current (mA crest)*	*Max. Disc. Temps. (°C)*	*5*	*10*	*15*	*20*
57	Polymer	81	<1	<38	35	42	44	44
66	Porcelain	54	68	>163	—	—	—	—
84	Polymer	109	<1	<38	50	52	60	60
98	Porcelain	122.4	18	<82	143	160	175	185

Tested using the 5-hour uniform slurry test procedure. This test consists of applying a uniform coating of standard 400 ohm-cm slurry to the test arrester. Within 30 seconds, MCOV is applied for 15 minutes, during which time the surface leakage currents cause the surface to dry. Slurry applications are repeated for a total of 20 test cycles. After the 20th test cycle, MCOV is applied to the arrester for 30 to 60 minutes to demonstrate thermal stability Surface leakage currents were measured at the end of the 5th, 10th, 15th and 20th test cycles.

Different polymer materials may be combined to produce a polymer with special properties; for example, a silicone EPDM is highly resistant to industrial type pollution and ocean salt.

The advantageous strength to weight ratio of polymer as compared to porcelain makes possible lighter structures and overall costs as well as permitting more compact designs, resulting in narrower right-of-way requirements and smaller station layouts. The reduction in handling, shipping, packaging, storage, preparation and assembly, all with less breakage, are obvious—these, in addition to the superior electromechanical performance.

Fiberglass insulation with its polymer petticoats is supplanting porcelain in bushings associated with transformers, voltage regulators, capacitors, switchgear, circuit breakers, bus supports, instrument transformers, lightning or surge arresters, and other applications. The metallic rod or conductor inside the bushing body may be inserted in a fiberglass tube and sealed to prevent moisture or contamination entering between the conductor and the fiberglass tube around which the polymer petticoats are attached. More often, the fiberglass insulation is molded around the conductor, and the polymer petticoats attached in a similar fashion as in insulators. Figure A-5.

Lightning or surge arrester elements are enclosed in an insulated casing. Under severe operating conditions, or as a result of multiple operations, the pressure generated within the casing may rise to the point where pressure relief ratings are exceeded. The arrester then may fail, with or

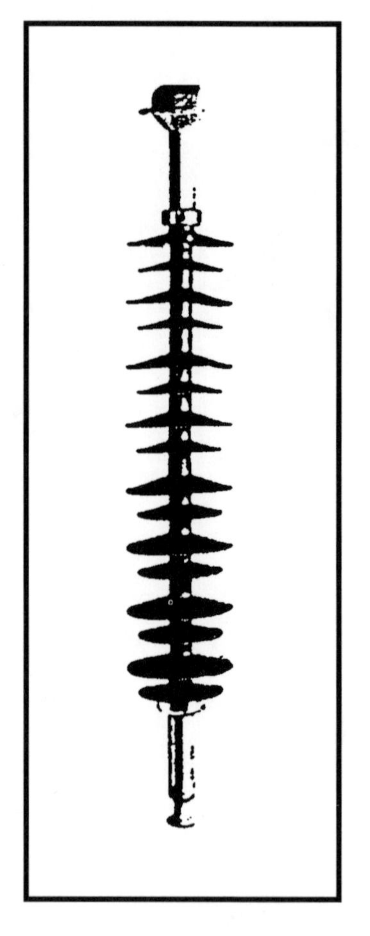

Figure A-4. Polymer insulator arrangement areas of high contamination where flashover between petticoat edges is possible *(Courtesy* Hubbe) Power Systems)

Figure A-5. Typical porcelain bushings that may be replaced with polymers. (a) Typical oil-filled bushing for 69 kV transformer.

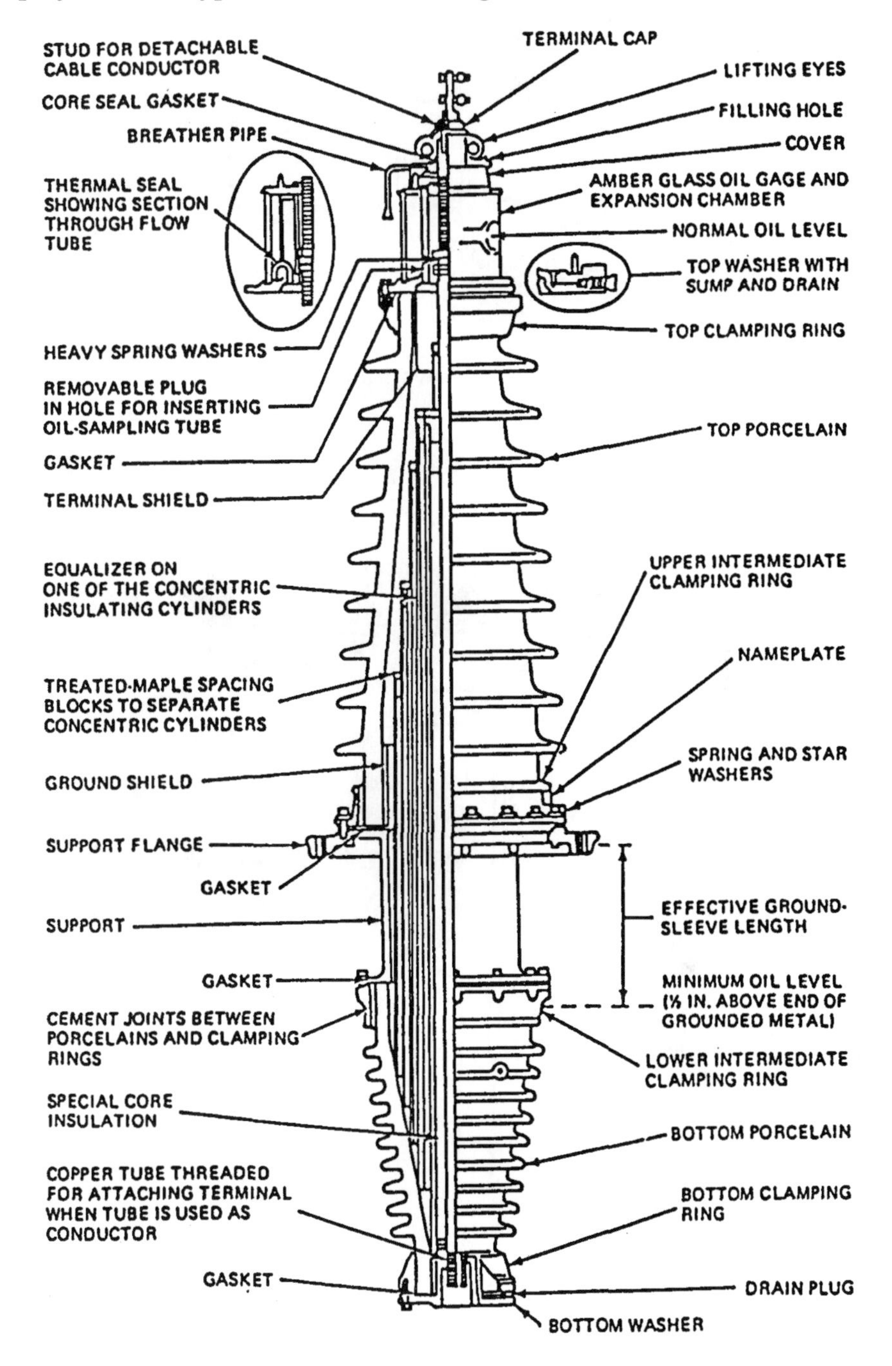

Figure A-5(b). Sidewall-mounted bushing.

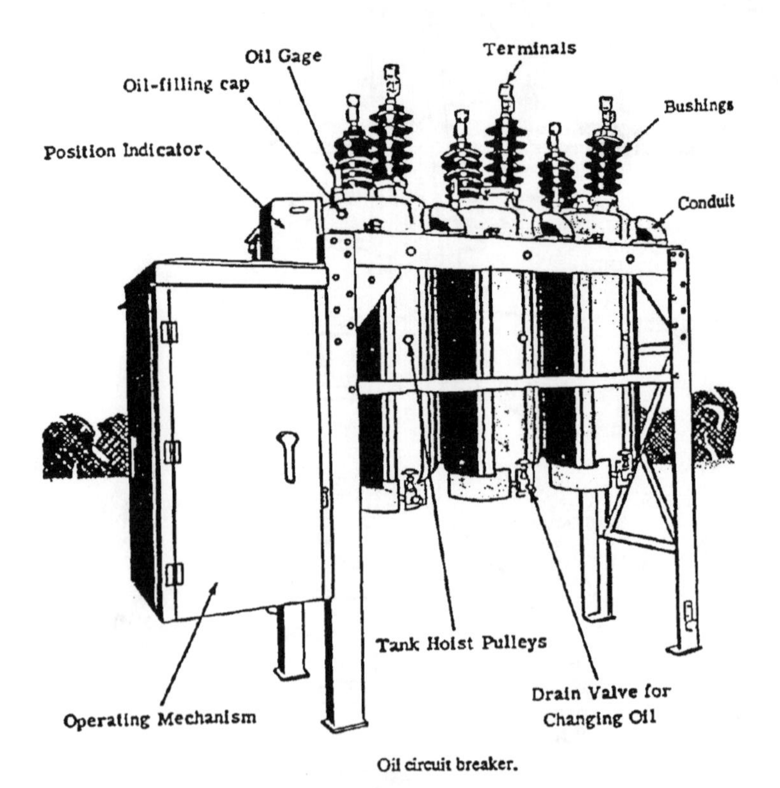

Oil circuit breaker.

Figure A-5(c).

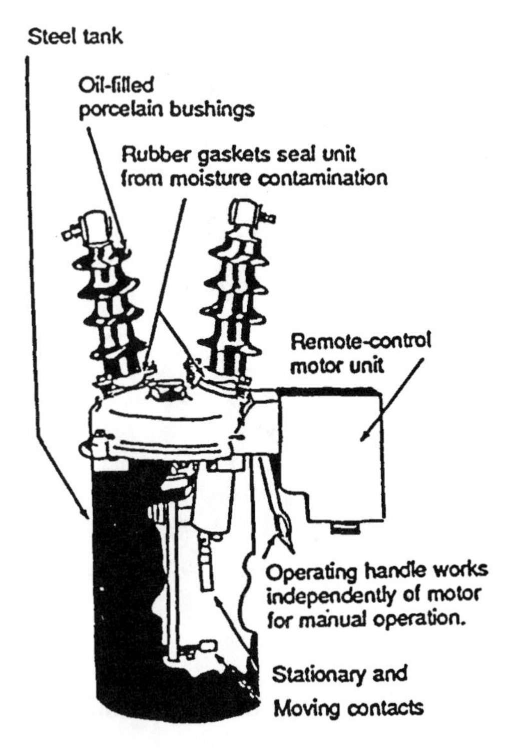

Figure A-5(d). Bushing applications that may be replaced with polymers.

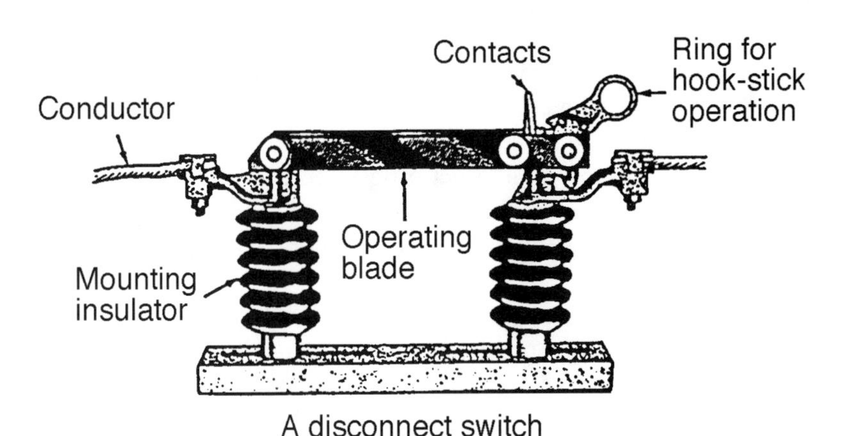

Figure A-5(e).

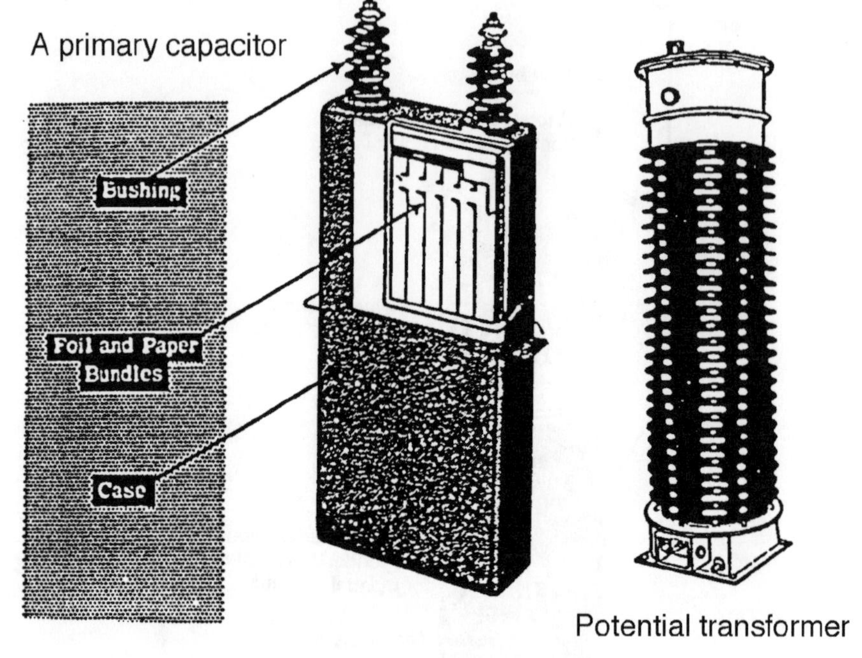

Figure A-5f. More porcelain insulation that may be replaced with polymers.

without external flashover, Figure A-6, exploding and violently expelling fragments of the casing as well as the internal components, causing possible injury to personnel and damage to surrounding structures. The action represents a race between pressures building up within the casing and an arcing or flashover outside the casing. The 'length' of the casing of the arrester limits its ability to vent safely. The use of polymer insulation for the casing permits puncturing to occur, without the fragmentation that may accompany breakdown and failure of porcelain.

Summarizing, the advantages of polymers over porcelain include:

- Polymer insulation offers benefits in shedding rain water or sleet, particularly in contaminated environments.

- Polymer products weigh significantly less than their porcelain counterparts, particularly line insulators, resulting in cost savings in structures, construction and installation costs. Table A-4.

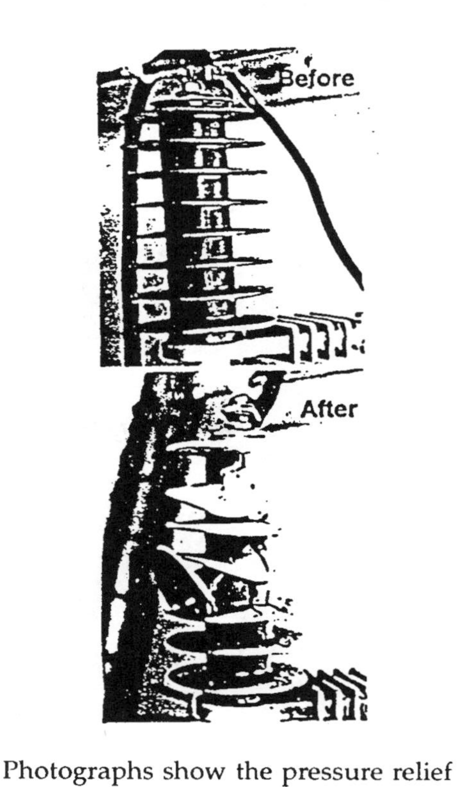

Photographs show the pressure relief capability of surge arresters.

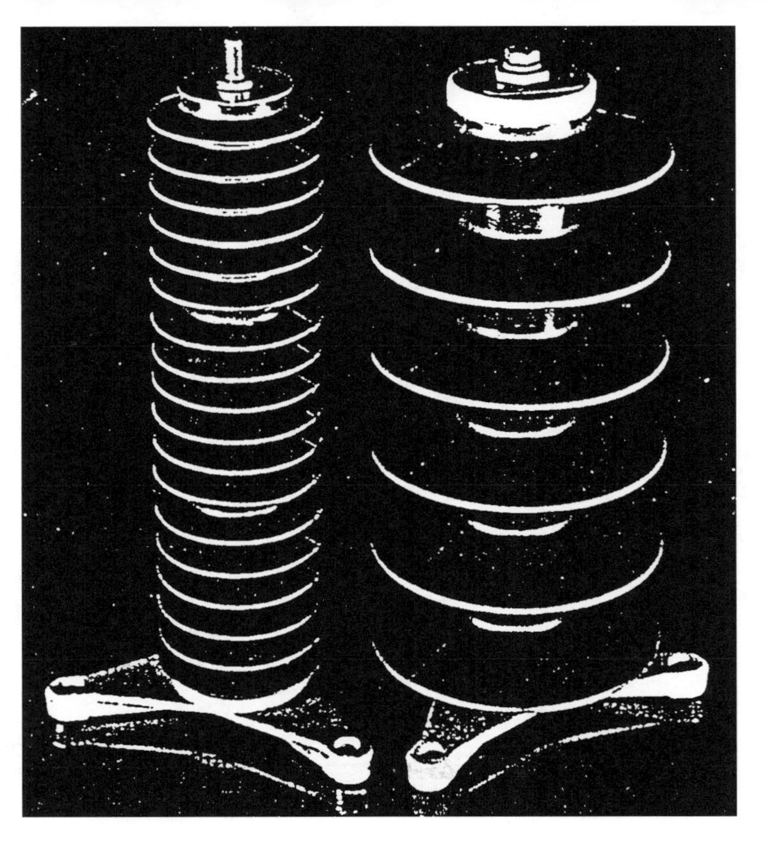

Polymer insulated arresters - station type

Figure A-6. Polymer and porcelain cased arresters.

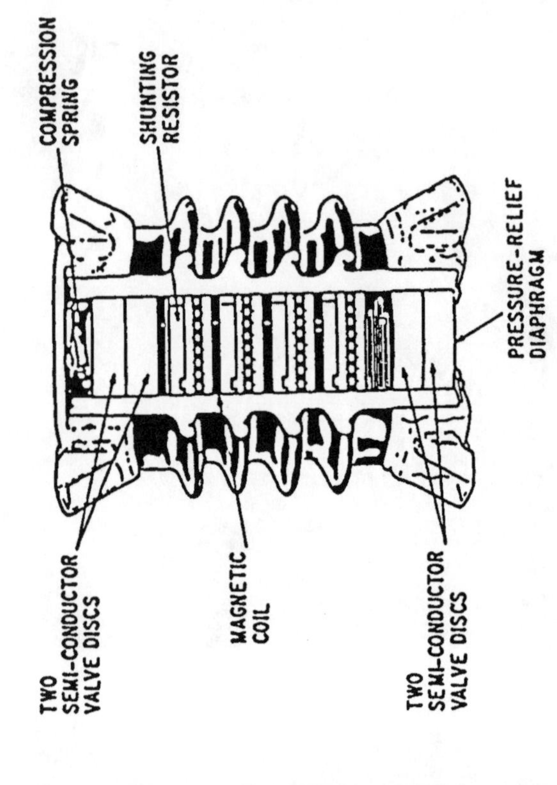

Cross-section of porcelain cased arrester

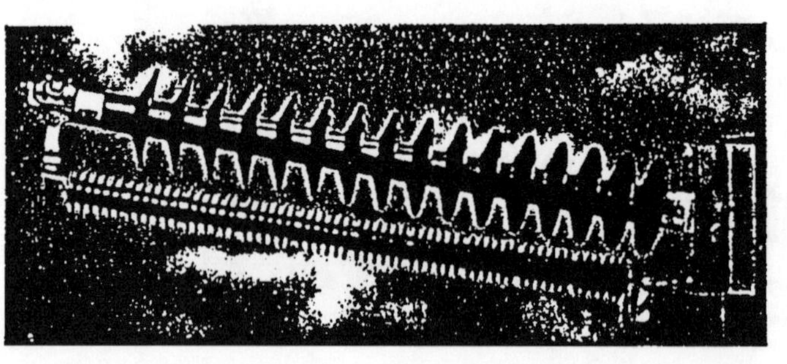

Polymer insulated intermediate type arrester

Figure A-6 (cont'd). Polymer and porcelain cased arresters.

Table A-4. Example—10 Miles, 345 kV, 250 Strings of Insulators

Porcelain -	4500 bells, 52-3, 13.5 lbs. ea., total 60,760 lbs.
	750 crates at 3.1 cu. ft. = 2,325 cu. ft.
	Insulator cost = $51.760 ($11.60/bell)
Polymers -	250 units, 14.4 lbs. ea., total 3 600 lbs.
	5 crates at 75 cu. ft. = 375 cu. ft.
	Insulator cost = $51.750

		Savings	
1.	Storage space at receiving point (3 mos.) porcelain - 580 sq. ft.; polymer - 100 sq. ft	480 sq. ft.	$ 60.00
2.	Off-load, re-load at receiving point; porcelain - 10 man-hrs.; polymer - 2 man-hrs.	8 man-hrs.	$120.00
3.	Breakage - off-loading, storage; re-load- porcelain - 1 percent; polymer - 0	1 percent	$517.50
4.	Truck - receiving point to tower sites (5 miles); porcelain- 1.00/cwt.; polymer .50/cwt	$589.50	$589.50
5.	Off load at tower site porcelain - 5 man-hrs.; polymer - 1 man-hr	4 man-hrs.	$60.00
6.	Unpack at tower site; porcelain - 50 crates/hour, 25 man-hrs.; polymer - 50 insulators/hour, 5 man-hrs	20 man-hrs.	$300.00
7.	Breakage - off-loading through string assembly & cleaning porcelain - 1 percent: polymer - 0	1 percent	$517.50
8.	Assemble strings, attach blocks porcelain - 40 man-hrs.; polymer - 8 man-hrs	32 man-hrs.	$480.00
9.	Clean insulators; porcelain-10 min./string; polymer - 3 min./string	29 man-hrs.	$435.00
10.	Lift string into place (2 men); porcelain - 5 min./string; polymer - 2 min./string	25 man-hrs.	$375.00
11.	Install & connect to tower (2 men); porcelain - 5 min./string polymer - 2 min./string	25 man-hrs.	$375.00
12.	Breakage - lifting & installation; porcelain - 0.5 percent; polymer - 0	0.5 percent	$258.75
13.	Cleanup packaging materials at jobsite; porcelain - 6 man-hrs.; polymer - 1.5 man-hrs	4.5 man-hrs.	$67.50

(*Courtesy* Hubbell Power Systems)

- Polymer insulators and surge arresters are resistant to damage resulting from installation and to damage from vandalism. The lack of flying fragments when a polymer insulator is shot deprives the vandal from his satisfaction with a spectacular event and should discourage insulators as convenient targets.

- Polymer arresters allow for multiple operations (such as may result from station circuit reclosings), without violently failing. Figure A-6.

- Polymer insulators permit increased conductor (and static wire) line tensions, resulting in lower construction designs by permitting longer spans, fewer towers or lower tower heights.

- Polymer one-piece insulators, lacking the flexibility of porcelain strings and the firm attachment of the conductor it support are said to produce a tendency to dampen galloping lines.

Although polymer insulation has become increasingly utilized over the past several decades, there are literally millions of porcelain insulated installations in this country alone; economics does not permit their wholesale replacement. Advantage is taken of maintenance and reconstruction of such facilities to make the change to polymers.

Much of the data and illustrations are courtesy of Hubbell Power systems, and is herewith duly acknowledged with thanks.

Appendix B

Street Lighting Constant Current Circuitry

STREET LIGHTING

Besides lighting homes, factories, and offices, it is the job of the utility company or local governing bodies to illuminate the streets and highways of the area it serves. In performing this job, as in every other, the aim is to supply the best possible service for the least possible expense.

The first and most important purpose of street lighting is safety-safety from traffic accidents and crime. Street lighting also adds considerably to the beauty of a town street or a highway. All in all, good street lighting spells comfort and convenience for the residents of a community (Figure B-1) or the users of highways.

The factors which the utility company must consider in planning a streetlighting system are many.

1. How much light is needed.
2. Types of structures along the road: homes or businesses.
3. Design of the road.
4. How much traffic passes by.
5. Crime potential.

Studies have proved conclusively that good street lighting considerably reduces traffic accidents and crime rates. It also improves retail business.

The Street-Lighting Fixture

The sight of a street or highway lighting fixture is certainly familiar. Generally, a street light consists of a post (wood, iron, aluminum, or

Figure B-1. The purpose of street lighting is safety from traffic accidents and crime.

concrete) (see Figure B-2), with an attached fixture which supports a lamp. Many times, the lamp is enclosed in diffusing glassware. This is to make for proper distribution of light. Here again, the utility company must consider safety. It is good lighting practice to prevent any glare effect. That's why these globes or glass-paneled fixtures are so important. The height and brilliancy of the unit, as well as the spacing between the units determine the intensity of illumination which will be produced. This will vary, of course, at different points between the lighting units. Many recent types of fixtures also contain a photocell, which is discussed later.

Street-Lighting Patterns

A street-lighting engineer can make light diffuse at any angle, depending on the area to be illuminated, the spacing between the units, and the mounting height of the luminaire. Although most of the light is directed to the road, there are many situations where the sidewalk must be comparably illuminated. The pendant (or hanging) type luminaire is most frequently used. The effectiveness of a pendant luminaire is considerable since it is usually hung right over the road.

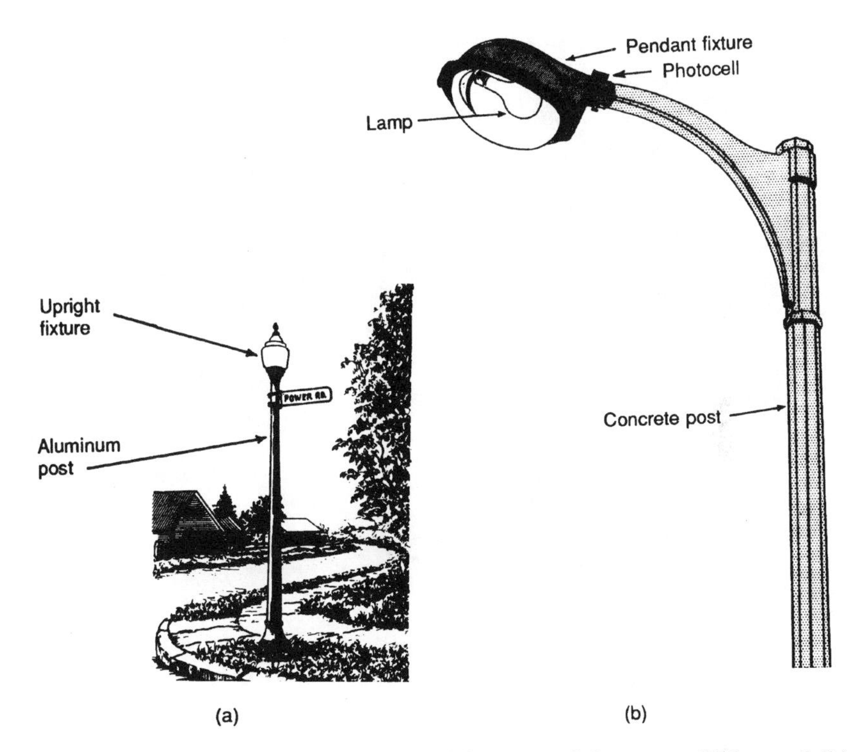

Figure B-2. Street-lighting fixtures. (a) An upright street light and (b) a pendant street light.

There are actually five different patterns of street-light diffusion shown in Figure B-3, differentiated by the degree of the diffusion angle. Type I consists simply of a light hung at an intersection. Type II is mounted at the side of a street and diffuses light at a 65 degree angle. Type III is also mounted on the side of a street; but in this case light is diffused at a 45 degree angle. In the asymmetric pattern (Type IV), light is diffused on both sides at a 90 degree angle and all the way across the street. The idea of the symmetric pattern (Type V) is to light up the whole intersection, roads, and corners alike.

MULTIPLE STREET LIGHTING

Street-Lighting Circuits

As will be explained again later in this book, the two fundamental

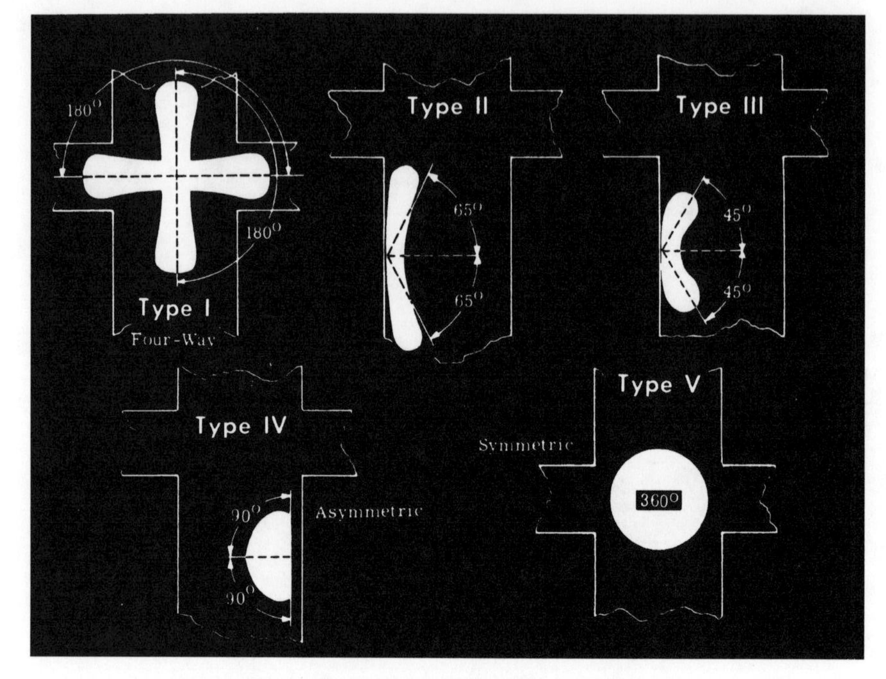

Figure B-3. Street-lighting patterns.

types of electric circuit are the *multiple* or parallel circuit and the *series* circuit. Other types are a combination of these. Both of these types of circuits are used for street lighting, and they are commonly referred to as *multiple lighting* and *series lighting*.

The series street-light system is rapidly being replaced by the multiple system with photoelectric control of individual or groups of street lights. However, as there are many of these series street-light systems still in existence, a description of this system will be included here.

Multiple Lighting

A parallel or multiple circuit may be likened to the rungs of a ladder as shown in Figure B-4(a). The rungs would represent the devices connected in the circuit and the sides represent the circuits applying a common pressure. Consider the water system shown in Figure B-4(b). The water is pumped from a source where its pressure is raised to give it the necessary pressure. When the valve is opened, the pump raises the pressure and sends it into the top pipe. When

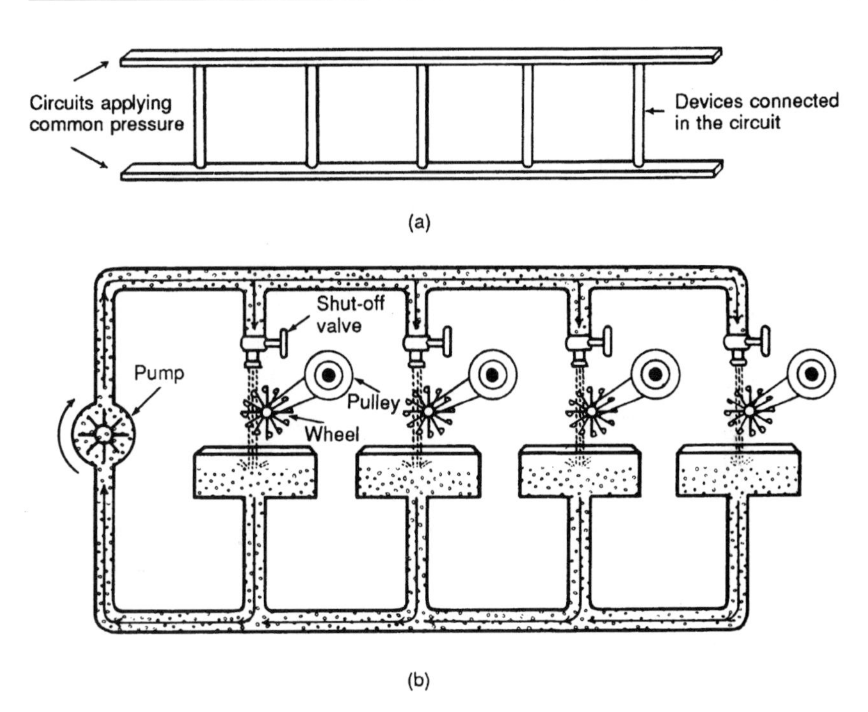

Figure B-4. (a) A multiple circuit is like a ladder. (b) Water analogy of a multiple circuit.

the valve associated with the first waterwheel is opened, the water under pressure turns the waterwheel which does some work. The water is collected and returned through the lower pipe, where it is returned to the pump, again raised in pressure, and so on around again.

The same action takes place when the valve associated with the second waterwheel is opened, the third waterwheel, and so on. In this type water system, the pressure applied to each of the several waterwheels is the same, but the current of water flowing through the pump is the sum of the currents flowing through each of the waterwheels.

Similarly, a multiple street-lighting system consists of a group of lamps in multiple connected to the secondary street mains through relays or time switches. The relay or time switch may be connected to control only the lamp at the post or pole on which it is located or it may control a number of lamps. Figure B-5 is a simplified diagram of a typical control arrangement.

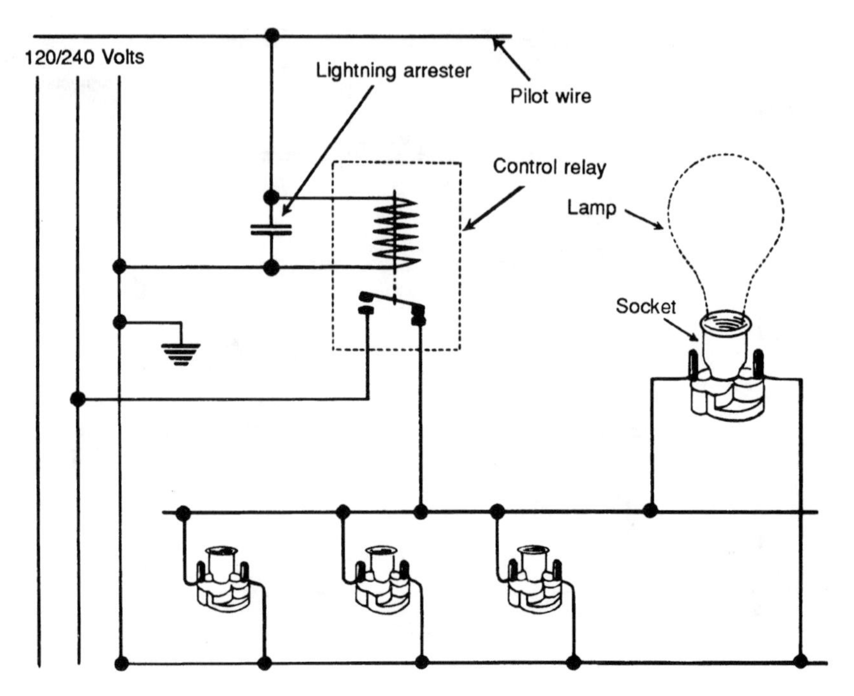

Figure B-5. Multiple street-lighting circuit showing control arrangement.

SERIES STREET LIGHTING

Series Lighting

To understand the series electric circuit, consider the water system shown in Figure B-6. The water is pumped from a source where its pressure is raised to give it the necessary force to travel around the system shown. When the valve is turned on, the water flows through the top pipe to spigot No. 1 where the outrushing water turns the waterwheel and does some work. As it falls into the collecting reservoir, the pressure of the water is reduced by the amount of pressure necessary to turn the first waterwheel. The remaining pressure pushes the water on to the second waterwheel where the same action is repeated. And again, the action is repeated to the third, and so on, until finally, the water pressure is all used up and the pressure drops to zero. The water then is collected in the bottom pipe, where the pump picks it up and the cycle is repeated. In this system, the pressure lost in turning each of the

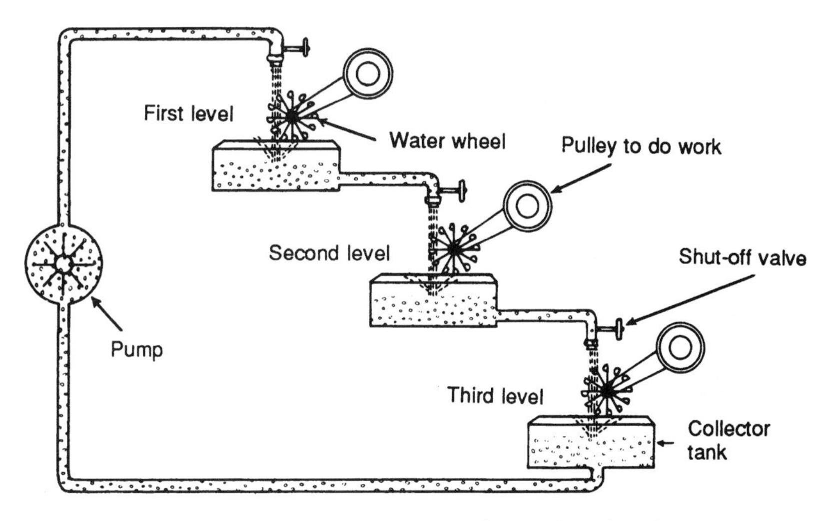

Figure B-6. Water analogy of a series circuit.

several waterwheels may be different, but the sum of these pressure-losses totals up to the pressure applied by the pump. The current of water is the same through each waterwheel.

Similarly, the series street-lighting circuit is operated by connecting the lamps in series. Since the lamps are connected in series, the same current passes through each lamp, even though the brilliance of light output for the various lamps may be different. It is obvious that the more lamps connected in the circuit, the greater the applied pressure or voltage will have to be to maintain a fixed current flowing through the circuit; similarly, the fewer lamps, the less the voltage.

All the parts of a series circuit are connected in succession, like beads on a string, so that whatever current passes through one of the parts passes through all of the parts. A typical series circuit is illustrated in Figure B-7.

The amount of light produced by such a series light depends on the length of the filament; the thickness for different light output lamps is the same. The greater amount of light required, the longer will be the filament. Lamps for series circuits are usually rated in *lumens* or "candle-power," that is units of light output rather than in units of electrical input. For larger size lamps, thicker size filaments are used which are designed for operation at 15 and 20 amperes. Ordinary lamps are designed for optimum operation when 6.6 amperes flow through them.

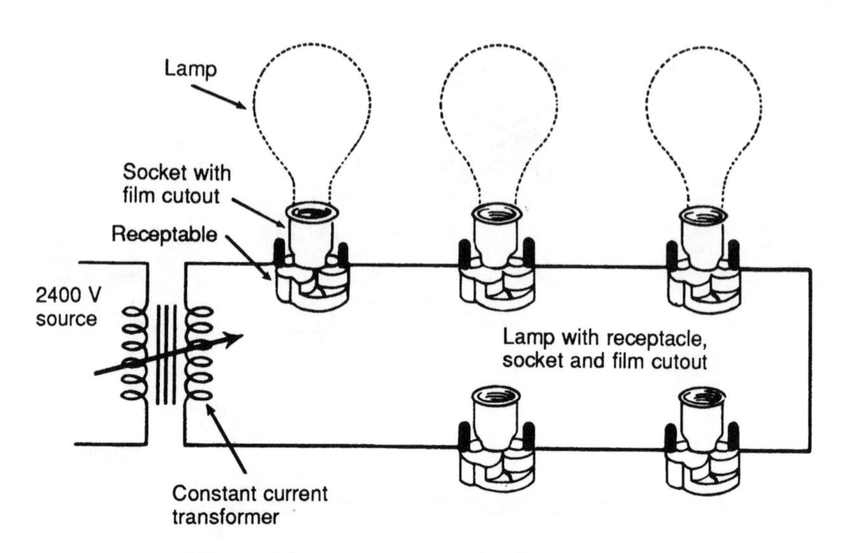

Figure B-7. Series street-lighting circuit.

Film Disc Cutout

In the water circuit illustrated, it should be noted that if there is a break in any of the connecting pipes, the water will drain out and the action stops. A closed valve causes the same halt in action.

Similarly, if one of the lamps in a series circuit were to burn out, the circuit would be interrupted and all the lights would go out. If the current were to be allowed to flow into a burned out lamp in a series circuit, it would be "dead-ended" there and not allowed to go any further. What is needed is a device to automatically "detour" the current on to the other lamps when one blows.

Such a device is known as a *film disc cutout* and is installed between the prongs of the series socket as shown in Figure B-8. It consists essentially of a thin piece of paper or insulation. Under normal conditions while current flows through the lamps, the pressure or voltage between the prongs of the lamp is insufficient to break down the piece of paper which is now acting as an insulator. However, when the lamp burns out, current can no longer flow and the pressure or voltage builds up on the prongs of the lamp, until the film of paper is punctured [see Figure B-8(b)]. Thus, the circuit is automatically closed, the burned out lamp being short-circuited and the rest of the lamps in the circuit remain lit.

A socket for series lighting is shown in Figure B-8(c). The prongs and the film cutout are indicated. For multiple lighting, a socket of the

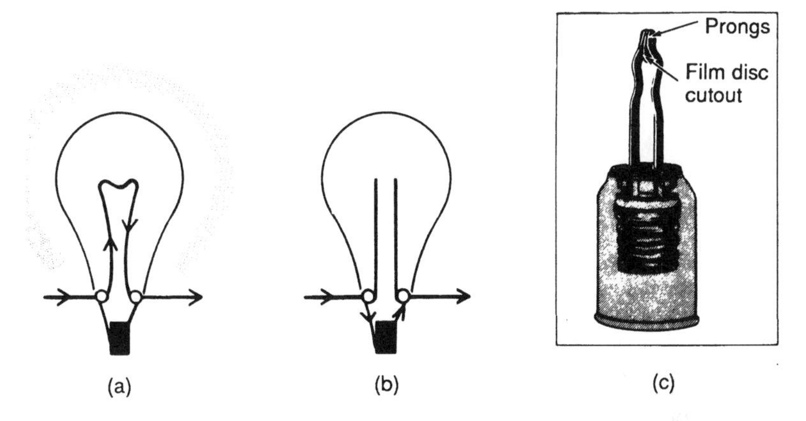

Figure B-8. Film disc cutout (a) acts as an insulator when lamp is lit, (b) is punctured and acts as a conductor when lamp burns out. (c) A series socket (phantom view) showing film disc cutout.

ordinary type, such as is found in household lamps is used. However, it may be larger in dimension than the household socket, and is then known as a Mogul socket.

The Constant-current Transformer

The lamps used in a series street-lighting circuit require a constant current. Most lamps are designed to carry a current of 6.6 amperes. Any change in this current value might cause flickering light, and therefore, produce poor illumination. As the number of lamps in a series circuit burn out and are then short circuited, the resistance of the circuit changes and the current in the circuit will fluctuate provided the applied voltage remains constant. Therefore, a good streetlighting circuit requires the use of a transformer which will automatically vary the voltage so that the current will remain constant.

Although many different mechanisms have been developed for the constant-current transformer, they all do essentially the same thing. The secondary coil moves up and down to vary the magnetic field.

The primary coil is usually fixed and the secondary floats (see Figure B-9). The latter is balanced by a weight which is proportional to the repulsion force of the rated current. If more than the rated secondary current flows, the repulsion becomes greater than the weight and pushes the secondary farther away from the primary. This automatically produces a greater leakage flux, which in turn, automati-

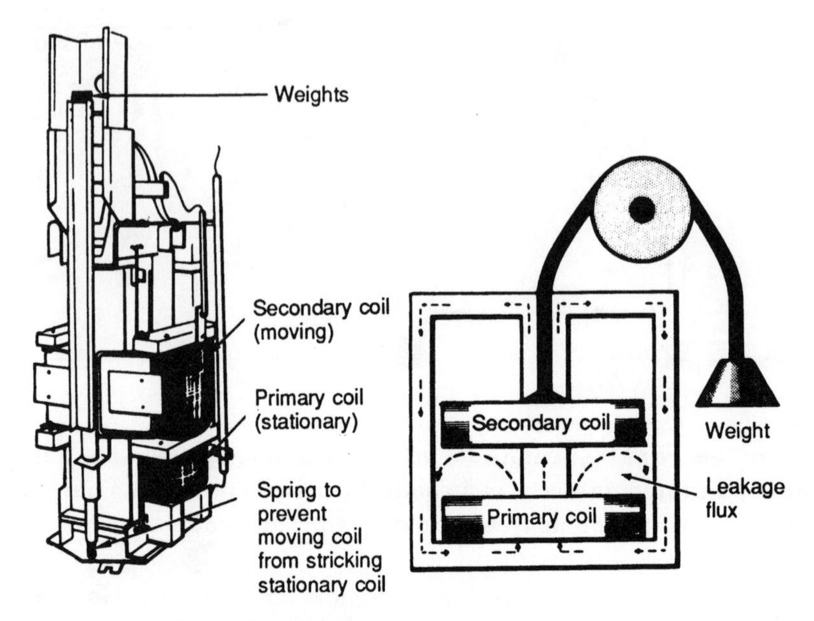

Figure B-9. Constant-current transformer.

cally lowers the voltage, thus automatically lowering the current to its rated value again.

CIRCUIT CONTROL

Astronomical Time Switch

The series circuit may be controlled at the substation. Here the circuit is turned on or off at the proper lighting times either by hand or by means of an *astronomical time switch* (see Figure B-10). This is merely a switch actuated by a relay which opens or closes its contacts in accordance with the hours of sunrise and sunset. These hours are reflected in the shape of a cam which is driven by a small electric motor. As it turns, it allows contacts to be made or broken, regulated by the shape of the cam. These may be replaced by photocell controlled relays similar to those located in the field.

The distinction between series and multiple lighting is that the series lamps are designed to operate on circuits regulated for constant *current* while multiple lamps are designed to operate on low-voltage circuits regulated for constant *voltage.*

Photocell-Controlled Relay

On and off control of street lights is often accomplished through the use of a photoelectric-controlled relay. Older systems use a phototube [Figure B-11 (a)], which consists basically of a pair of electrodes contained in an evacuated envelope. When properly connected into a circuit, these electrodes allow more current to flow in the tube as the intensity of the incident light increases. The phototube circuit is integrated with a control circuit which incorporates a relay. When the current in the phototube circuit reaches a predetermined value, the relay in the control circuit functions and causes the lights to go out. As the day darkens, the reverse action takes place. The current in the phototube circuit decreases until another predetermined value is reached. Then the relay in the control circuit functions and the lights are turned on.

Newer systems use a solid-state photocell [Figure B-11 (b)] instead of a phototube. The operation is essentially the same as that of the pho-

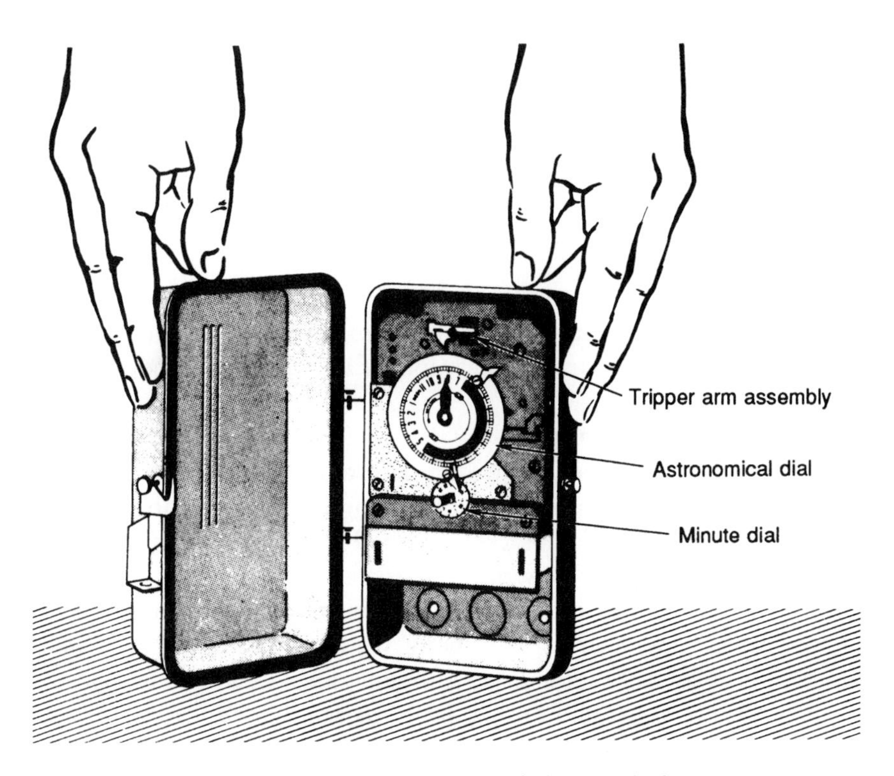

Figure B-10. Astronomical time switch.

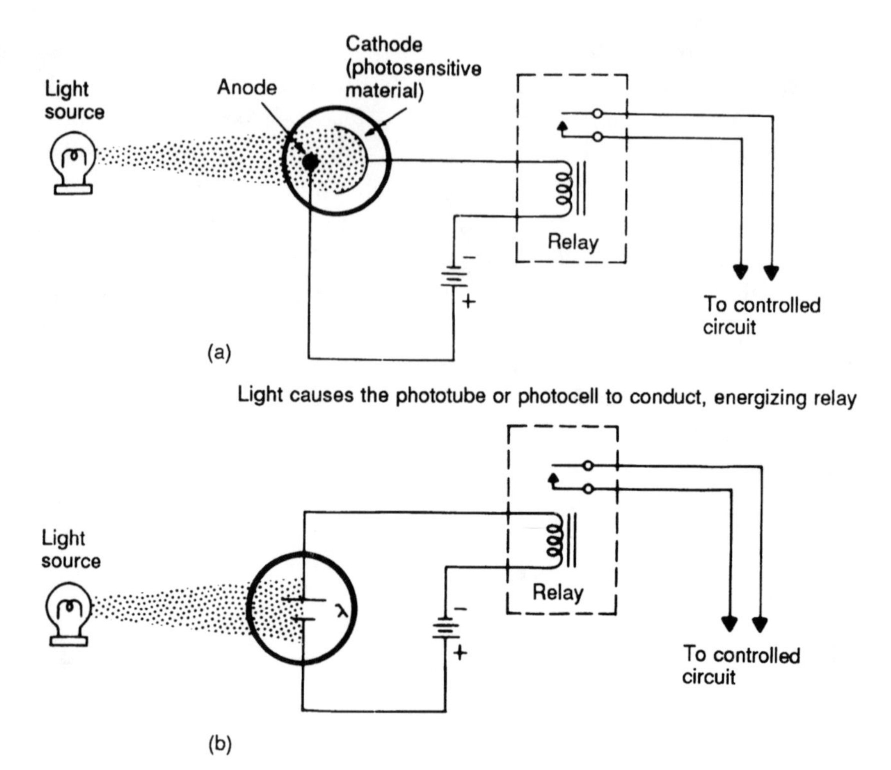

Figure B-11. Photoelectric-controlled relays. (a) Phototube and (b) photocell.

totube system, but the use of a photocell reduces maintenance because the life of a photocell is very much longer than that of a phototube.

Street-Lighting Accessories

Street lights may be installed on steel, aluminum, or concrete standards, usually fed from underground cables. More often they are attached to some fixture or bracket attached to a pole. The shapes and types of fixtures vary widely.

Street-light lamps, without any equipment, in general emit light in all directions. Upward light may be of value in certain business areas where the facades of buildings so illuminated act as secondary sources of light, reduce glare, and are generally pleasing in appearance. However, in most cases it is desirable to modify the distribution of light from the lamps and three general types of equipment are

1. reflectors,
2. refractors,
3. diffusing glassware.

Lamps may be equipped with reflectors only, refractors only, or diffusing globes only. They may also be equipped with any combination of these three or with all three. The three devices are sometimes incorporated into one unit (see Figure B-12).

A reflector (Figure B-13) is practically opaque; the light rays hitting its surface are stopped and redirected in the desired direction. A reflector may be of porcelain, enameled porcelain, metal, silvered glass, and silvered or other plated metals. There are various shapes of reflectors designed to give different types of light distribution.

A refractor (Figure B-14) is transparent; the light rays hitting its surface are bent and pass through in the desired direction. In the ordinary lamp, a large portion of the light is emitted in directions away from the street, and is wasted in so far as street illumination is concerned. It is the function of the refractor to intercept a large share of this light and

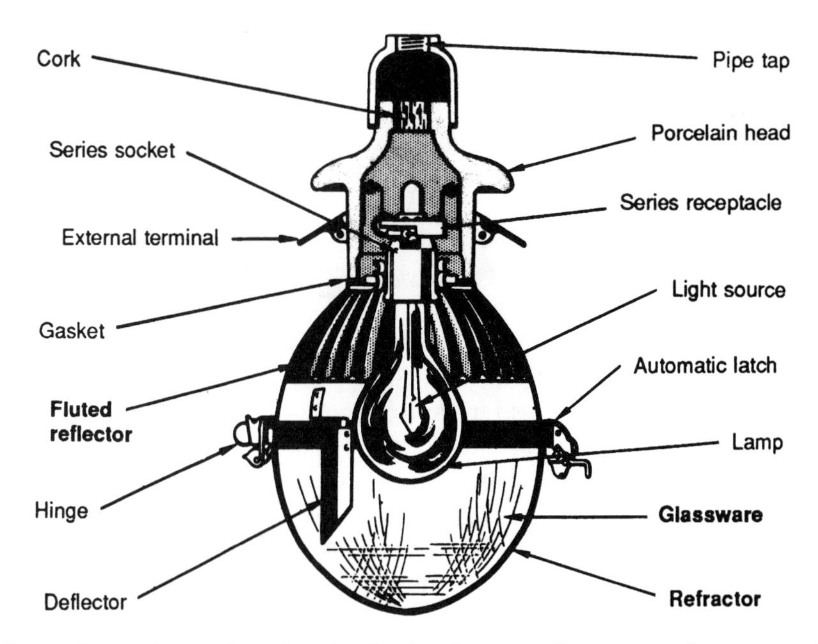

Figure B-12. A pendant luminaire having a reflector, a refractor, and diffusing glassware.

to redirect it on the street surface. This builds up the illumination around the street light fixture and utilizes the light more efficiently.

Glassware is used for various purposes. It may be used for protection to the lamp, or it may be used, as in most cases, in order to decrease the brilliancy of the light source and make the globe itself act as a secondary light source. In performing the latter function, diffusing glass is used (Figure B-15). This glass is usually *rippled* in some way, to diffuse the light and make a large, even light source. To reduce damage from vandalism, glass is often replaced by clear plastic.

Figure B-13. A fluted reflector.

STREET-LIGHTING LAMPS

Street-lighting Lamps

The usual lamp used in street-lighting circuits is of the regular incandescent type similar to those found in the home. In this type, light is given off from a piece of wire (filament) heated white hot. The amount of light which can be obtained from this type lamp is limited by its physical dimensions and the manufacturing facilities.

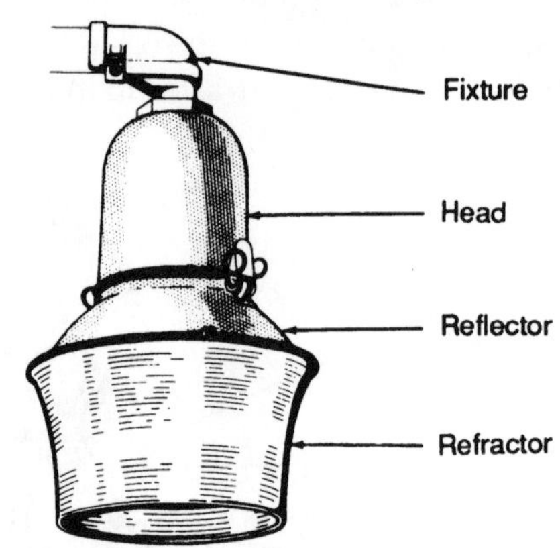

Figure B-14. One type of refractor.

The important elements of the incandescent lamp shown in Figure B-16, (or bulb, as we call it in the home) are the tungsten filament, a mandrel on which this filament is wound, and inert gases which cool the filament to cut down on evaporation and carry any tungsten which does evaporate to the top of the lamp. The reader has often noticed that an incandescent bulb gets a dark spot on top after long use. This is the evaporated tungsten. The gas is usually a nitro-

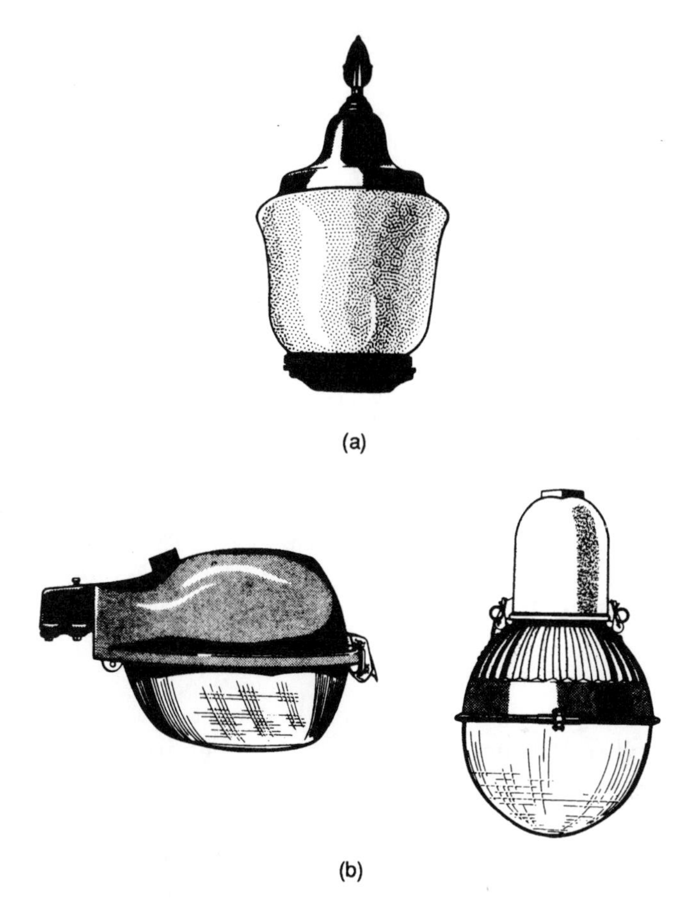

Figure B-15. Diffusing glassware prevents glare, is usually rippled, and comes in many different shapes. (a) Upright fixture (underground wired) and (b) pendant fixtures (overhead wired).

gen-argon combination.

To obtain more illumination from devices and still maintain practical dimensions, other types of lighting have been developed. Fluorescent lamps shown in Figure B-17(a) usually one or more tubes 4 feet in length, similar to those found in stores, factories, and other establishments, are used along boulevards, in tunnels, parking lots and other special situations where brilliant lighting is required. These lamps are connected for multiple circuits. A coating of fluorescent material applied to the inside of the glass tube is made to glow and give off light when acted on by rays, somewhat similar to x-rays, pro-

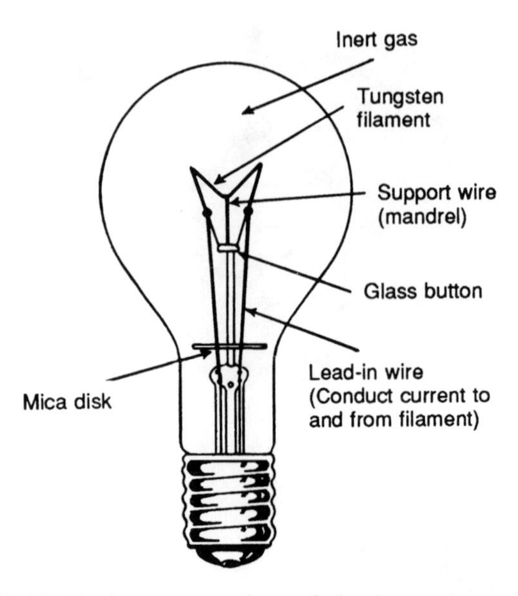

Figure B-16. Basic construction of the incandescent lamp,

duced by the flow of electricity.

The fluorescent lamp unit generally consists of 2 electrodes, one at either end of the tube. These electrodes emit and receive, receive and emit electrons between them in an operation that resembles a tennis match. A drop of mercury [see Figure B-17(b)] is added inside the tube to make it start or light up initially. A "ballast" device is included in the circuit to permit the electrodes to reach sufficient voltage to start the arc going across the tube. Fluorescent tubes can be made to shed other colors besides the well known blue-white.

Discharge Lamps

Where high-intensity illumination is required (along highways, for example) a discharge lamp fits the bill. The primary advantage of a gaseous discharge lamp is that it can give out much more light for its size than an incandescent lamp of comparable size. Sodium and mercury are commonly used materials for discharge lamps (see Figure B-18). Designers have discovered that all-around these are the most economical chemicals with respect to supplying the proper temperature, voltage, and pressure.

In the mercury-vapor lamp, the mercury is vaporized by the current flowing through the lamp. (Again the current is controlled by

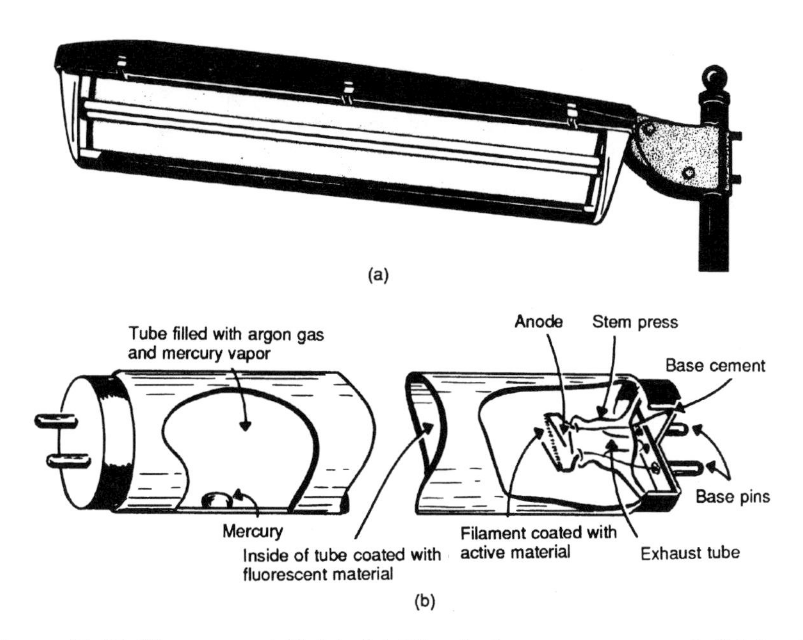

Figure B-17. Fluorescent light, (a) Physical appearance and (b) basic construction.

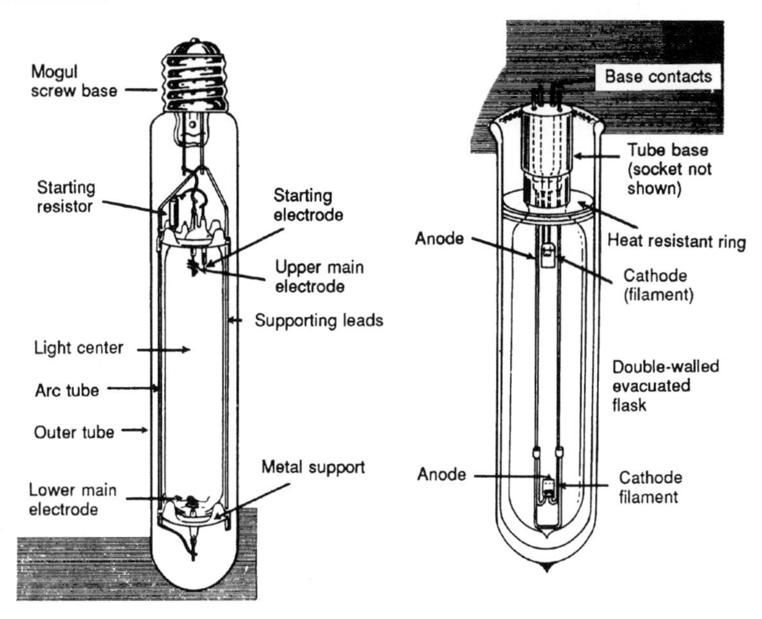

Figure B-18. Discharge lamps. The sodium-vapor lamp gives off a yellow-orange light. The mercury-vapor lamp gives off a green-blue light.

a ballast.) When the current continues through the gaseous mercury, the gas gives off great amounts of concentrated green-blue light.

With sodium, the principle of operation is the same, but the light produced is a highly intense yellow-orange. Discharge lamps are usually designed for multiple-circuit operation. Other materials are also used and are generally included in the class of halogen lamps.

Appendix C

The Grid Coordinate System: Tying Maps to Computers*

Anthony J. Pansini, E.E., P.E.

(The quality of electric service is, importantly, determined by the rapidity of restoration to normal during contingency conditions. This consideration is intimately tied to economics that, in turn, affects the design, construction, and operation of utility systems. Here, the ability to locate facilities and obtain data constitutes perhaps the greatest contribution to the quality of service.)

INTRODUCTION

The grid coordinate system is the key that ties together two important tools, maps and computers. Maps are a necessity for the better operation of many enterprises, especially of utility systems. Their effectiveness can be increased manyfold by adding to their information data contained in other files. Much of the latter data are now organized and stored in computer-oriented files-on punched cards and on magnetic tapes, drums, disks, and cells. Generally, these data can be retrieved almost instantly by CRTs (cathode ray tubes) or printouts. The link that makes the correlation of data contained on the maps and in the files practical is the grid coordinate system.

Essentially, the grid coordinate system divides any particular area served into any number of small areas in a grid pattern. By superimpos-

*Reprinted (with modifications) from Consulting Engineer, 8 January 1975, vol. 44, no. 1, pp. 51-55.

ing on a map a system of grid lines, and assigning numbers to each of the vertical and horizontal spacings, it is possible to define any of the small areas by two simple numbers. These numbers are not selected at random, but have some meaning. Like any graph, these two coordinates represent measurements from a reference point; in this respect they are similar to navigation's latitude and longitude measurements.

Further subdivision of the basic grid areas into a series of smaller grids is possible, each having a decimal relation with the previous one (i.e., by dividing each horizontal and vertical space into tenths, each resultant area will be one hundredth of the area considered). By using more detailed maps of smaller scale, it is possible to define smaller and smaller areas simply by carrying out the coordinate numbers to further decimals. For practical purposes, each of these grid areas should measure perhaps not more than 25 feet by 25 feet (preferably less, say 10 feet by 10 feet) and should be identified by a numeral of some 6 to 12 digits.

For example, by dividing by 10, an area of 1,000,000 feet by 1,000,000 feet (equivalent to some 190 square miles) can be divided into 100 smaller areas of 100,000 feet by 100,000 feet each, identified by two digits, one horizontal and one vertical. This smaller area can again be subdivided into 100 smaller areas of 10,000 feet by 10,000 feet each, identified by two more digits, or a total or four with reference to the basic 1,000,000 feet square area. Breaking down further into 1000- by 1000 feet squares and repeating the process allows these new grids to be identified by two more digits, or a total or six. Again dividing by 10 into units of 100 feet by 100 feet, and adding two more digits, produces a total of eight digits to identify this grid size. One more division produces grids of 10 feet by 10 feet and two more digits in the identifying number-for a total of 10 digits, not an excessive number to be handled for the grid size under consideration; see Figure A-1.

This process may be carried further where applications requiring smaller areas are desirable; however, each further breakdown not only reduces the accuracy of the measurements, but also adds to the number of digits, which soon becomes unmanageable. Experience indicates that a "comfortable" system should contain 10 digits or fewer for normal usages.

While the decimal relation has been mentioned, other relations can be used, such as sixths, eighths, and so on, or combinations, such as eighths and tenths, and others.

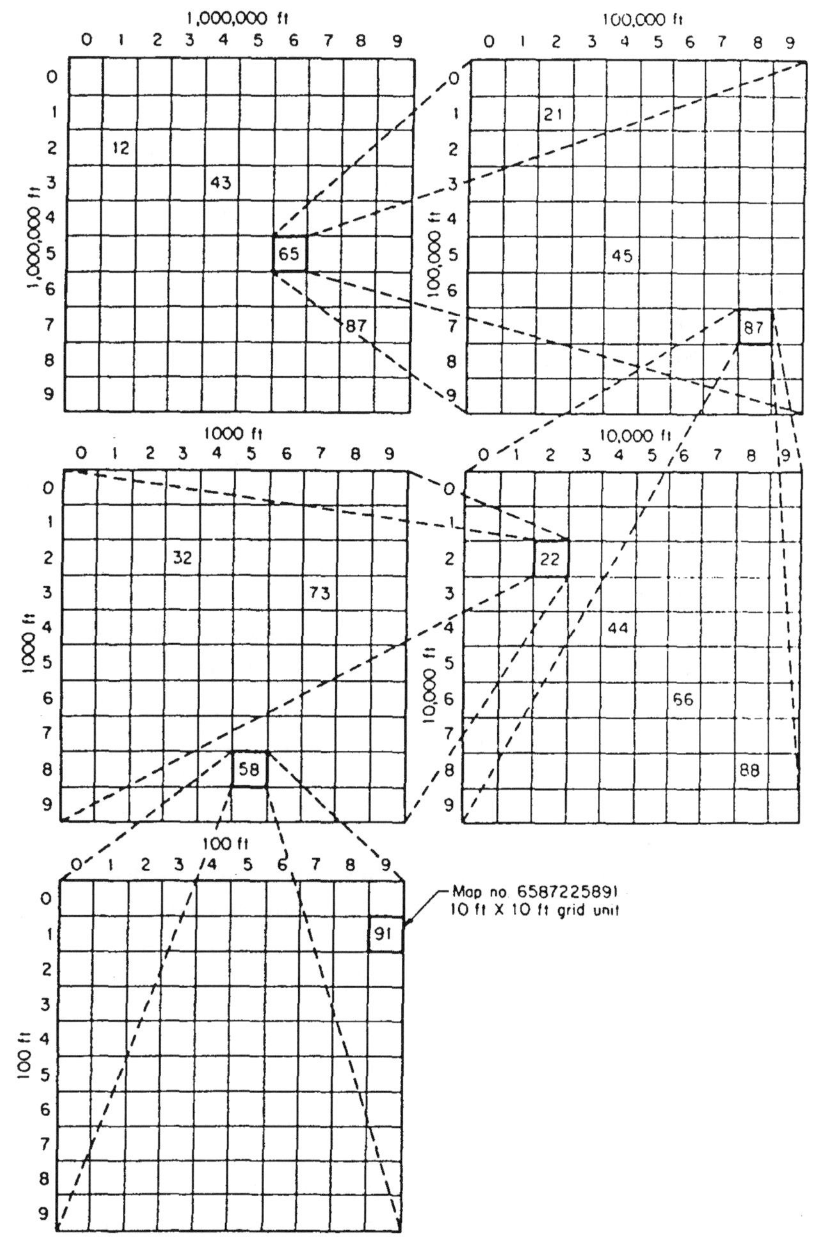

Figure C-1. Development of a grid coordinate system.

Standard References

The grid pattern applicable to each of the several scale maps may be printed directly on each map as light background lines, perhaps even in a different color, or printed on the back of the maps when they are reproduced. Alternately, a grid overlay can be applied to each map to be used when it is necessary to determine a grid coordinate for an item on the map. The actual grid number need not be printed on every item on every map unless desired. Such numbers assigned to key locations on each map normally suffice; the others may be determined from the grid background or overlay. To maintain their permanence and to minimize distortion from expansion and contraction because of changing humidity and temperature, the maps should be printed on a material such as Mylar, a translucent polyester-base plastic film; this is especially true of the base maps from which others of different scales and purposes are derived.

As much as practical, the data on the maps should be uncluttered and as legible as possible. It may be desirable in some instances to provide two or more maps (of the same scale) for several purposes; marks for coordinating these several maps, should it be necessary, may be included on each of the maps.

Maps may be further uncluttered by deliberately removing as much of the information on them as appears desirable and practical and consigning such information to files readily accessible by computer. In many instances, this information already is included or duplicated in such files, but may need to be labeled with the appropriate grid coordinate number. The use of CRTs and printouts makes this information available at will.

The grid coordinate number corresponding to the location of an item in question is added to its record and now becomes its computer address.

COORDINATE DATA HANDLING

As implied earlier, the grid coordinate system provides an easy and simple but, more important, a very rapid means of obtaining data from files through the use of the computer. In some respects, it assigns addresses to data in the same way as the ZIP code system in use by the postal service. The manner in which the grid number may be used is illustrated in the following examples; for convenience they refer to electric utility systems, although obviously they apply equally well to other endeavors employing maps and records.

Data contained on maps and records generally apply to the consumers served and the facilities installed to serve them. While maps depict (by area) the geographic and functional (electrical) interrelationship between these several components, the records supply a continuing history (by location) of each component item (consumers and facilities).

In the case of consumers, such data may include, in addition to the grid coordinate number, the name and post office address. Also a history of electric consumption (and demand where applicable), billing, and other pertinent data over a continuing period, usually 18 or 24 months. There may also be data on the consumer's major appliances; also the data and work order number of original connection and subsequent changes. The grid coördinate number of the transformer from which the consumer is supplied is included, as well as that for the pole or underground facility from which the service to the consumer is taken. Sometimes interruption data may be included. Other data may include telephone number, tax district, access details, hazards (including animals), dates of connection or reconnection, insurance claims, easements, meter data, meter reading route, test data, credit rating, and other pertinent information. Only a small portion of these data are shown on maps, usually in the form of symbols or code letters and numerals. In the case of facilities, such data may include, in addition to the grid coordinate number, location information, size and kind of facility (e.g., pole, wire, transformer, etc.). date installed or changed, repairs or replacements made (including reason therefor, usually coded), original cost, work order numbers, crew or personnel doing work, construction standard reference, accident reports, insurance claims, operating record, test data, tax district, and other pertinent information. Similarly, only a small portion of these data are shown on maps, usually in the form of symbols or code letters and numerals.

Data from other sources also may be filed by grid number for correlation with consumer and facility information for a variety of purposes. Such data may include government census data; police records of crime, accidents, and vandalism; fire and health records; pollution measurements; public planning; construction and rehabilitation plans; zoning restrictions; rights-of-way and easement locations; legal data; plat and survey data; tax district; and such other information that may affect or be useful in carrying out utility operations.

Obviously, all these data, whether pertaining to the consumer or to the utility's facilities, are not necessarily contained on one map or in one record only; indeed, there may be several maps and records involved, each containing certain amounts of specialized or functionally related data. All, however, may be correlated through the grid coordinate system.

Data Retrieval

Data contained in the files may be retrieved by means of the computer and may be presented visually by means of CRTs for one-time instant use, or by printouts and automatic plotting for repeated use over an indefinite time period. Data presented may be the exact original data as contained in one or more files, or extracted data obtained as a result of correlating data residing in one or more files, or a combination of both; such extracted data may or may not be retained in separate files for future use.

These data may be retrieved for an individual consumer or an individual item of plant facilities, or may be other data for a particular area, small or large. The various specific purposes determine what data are to be retrieved and how they are to be presented. They also determine the programs and equipment required. Data thus retrieved then are used with data contained on the map to help in forming the decisions required. The decisions may include new data that can be reentered in the files as updating material, that can be plotted or printed for exhibit purposes, or that can be re-entered on maps for updating or expanding the material thereon; all of these may be done by means of the computer.

The grid coordinate number is applied to utility facilities for ease of location and positive identification in the field. In the case of electric utilities, these may include services, meters, poles, towers, manholes, pull boxes, transformers, transformer enclosures, switches, disconnects, fuses, lightning arresters, capacitors, regulators, boosters, streetlights, air pollution analyzers, and other equipment and apparatus; also the location of laterals on the transmission and distribution circuits.

OTHER APPLICATIONS

Similarly, for gas utilities, the applications of grid coordinate numbers may include mains, services, meters, regulators, valves, sumps, test pits, and other equipment; also the location of boosters, laterals, and nodes on the gas systems. For water systems, they may include mains, services, meters, valves, dams, weirs, pumps, irrigation channels, and other facilities. For telephone and telegraph communication systems, including CATV circuits, they may include mains, services, terminals, repeaters, microwave reflectors, and other items including poles, manholes, and special items.

Grid coordinate numbers also may find application in many other lines of endeavor; highway systems, railway systems, oil fields, social surveys (police, health, income, population distribution, etc.), market surveys (banks and industries), municipal planning and land use studies, nonclassical archaeology, geophysical studies, and others where such means of location identification may prove practical.

The use of grid coordinates facilitates positive identification in the field; the numbers are posted systematically on facilities, such as streetlight or traffic standards, poles, and structures, and at corners or other prominent locations.

An atlas, consisting of a grid overlay on a geographical map, aids the field forces in locating consumers and plant facilities and provides a common basis for communication between office and field operating personnel.

With the national consensus apparently pointing to an ultimate metric system for the United States for conform with world standards, the adoption of a grid coordinate system provides an excellent opportunity for its introduction with a minimum of conversion effort.

With the advent of the computer, it was inevitable that the grid coordinate system should be developed to provide a simple means of addressing the computer. The grid number provides the link between the map and the vast amount of data managed by the computer. This happy marriage of two powerful tools results not only in better operations but in improved economy as well. It is a must in the modernization of operations in many enterprises and especially in utility systems.

Appendix D

United States and Metric Relationships

U.S. to Metric		Metric to U.S.	
Length			
1 inch	= 25.4 rum	1 millimetre	= 0.03937 inch
1 inch	= 2.54 cm	1 centimeter	= 0.3937 inch
1 inch	= 0.0254 m	1 metre	= 39.37 inch
1 foot	= 0.3048 m	1 metre	= 3.2808 feet
1 yard	= 0.9144 m	1 metre	= 1.094 yard
1 mile	= 1.609 km	1 kilometre,	= 0.6214 mile
Surface			
1 inch^2	= 645.2 mm^2	1 millimetre^2	= 0.00155 inch^2
1 inch^2	= 6.452 cm^2	1 centimetre^2	= 0. 155 inch^2
1 foot^2	= 0.0929 m^2	1 metre^2	= 10.764 foot^2
1 yard^2	= 0.8361 m^2	1 metre^2	= 1. 196 yard ^{2}
1 acre	= 0.4047 hectare	1 hectare	= 2.471 acres
1 mile^2	= 258.99 hectare	1 hectare	= 0.00386 mi^2
1 mile^2	= 2.59 km^2	1 kilometre^2	= 0.3861 mile^2
Volume			
1 inch^3	= 16.39 cm^3	1 centimetre^3	= 0.061 inch ^{3}
1 foot^3	= 0.0283 m^3	1 metre^3	= 35.314 foot^3
1 yard^3	= 0.7645 m^3	1 metre^3	= 1.308 yard^3
1 foot^3	= 28.32 litres	1 litre	= 0.0353 foot^3
1 inch^3	= 0.0164 litre	1 litre	= 61.023 inch^3

U.S. to Metric		Metric to U.S.	
Volume (continued)			
1 quart	= 0.9463 litre	1 litre	= 1.0567 quarts
1 gallon	= 3.7854 litres	1 litre	= 0.2642 gallons
1 gallon	= 0.0038 m^3	1 metre ^{3}	= 264.17 gallons
Weight			
1 ounce	= 28.35 grams	1 gram	= 0.0353 ounce
1 pound	= 0.4536 kg	1 kilogram	= 2.2046 lb*
1 net ton	= 0.9072 T (metric)	1 Ton (metric)	= 1.1023 net tons**
Compound units			
1 lb/ft	= 1.4882 kg/m	1 kilogram/metre	= 0.6720 lb/ft
1 lb/in^2	= 0.0703 kg/m^2	1 k g/cm^2	= 14.223 lb/in^2
1 lb/ft^2	= 4.8825 kg/m^2	1 kg/m^2	= 0.2048 lb/ft^2
1 lb/ft^3	= 16.0192 kg/m^3	1 kg/m^3	= 0.0624 lb/ft^3
1 ft-lb	= 0. 1383 kg-m	1 kg-m	= 7.233 ft-lbs
1 hp	= 0.746 kW	1 kW	= 1.340 hp
1 ft-lb/in^2	= 0.0215 kg-m/cm^2	1 kg-cm/m^2	= 46.58 ft-lb/in^2
Temperature			
1 degree F	= 5/9 degree C	1 degree C	= 9/5 degree F
Temp °F	= 9/5°C + 32	Temp °C	= 5/9(°F – 32)

*Avoirdupois

**1 ton = 2000 lb

Index